Mysteries of Electricity Definitively Solved and Simply Explained

Part 2:
Creation and Emission of Electromagnetic Energy

by Mark Fennell
© 2014

Websites of Mark Fennell

1. <u>Energy Technologies Explained Simply</u>
   http://EnergyTechnologySimply.blogspot.com

2. <u>Mark Fennell: Authentic Expressions of a Happy Dancer and Multidimensional Intellectual</u>
   http://markfennell.blogspot.com

3. <u>You Tube Channel: All Things Energy</u>
   http://www.youtube.com/channel/UCk5ckPqF4oD0JoJMSBi2Zcg

Electromagnetic Energy Books by Mark Fennell

- Mysteries of EM Energy: Definitively Solved and Simply Explained
  - Part 1: Fundamental Properties of Electromagnetic Energy

Energy Books by Mark Fennell

<u>Renewable Energy</u>
- Introduction to Electrical Power
- Wind Power Technology Explained Simply
- Hydropower Explained Simply
- Solar Power Technologies Explained Simply
- Practical Considerations of Solar Power
- Advanced Solar Cell Technologies

<u>Natural Gas, Coal, and Other Hydrocarbon Books</u>
- Natural Gas and Other Hydrocarbon Technologies Explained Simply
- Extracting and Refining Natural Gas (includes Fracking)
- Transportation, Storage, and Use of Natural Gas
- Coal Power Technologies Explained Simply
- Clean Coal Technologies
- Mercury and Coal Power

<u>Nuclear Power Books</u>
- Nuclear Power Technologies Explained Simply
- Health Hazards of Radioactive Decay
- Radiation Measurements
- Processes of Radioactive Decay and Storage of Nuclear Waste

<u>Power Line and Grid Books</u>
- Transmission of Electrical Power Explained Simply
- Utility Operations and Grid Systems Explained Simply

© Copyright Notice ©

This work is copyrighted to the author. No part of this book can be published or presented to the public as the work of any other author.

Sections of this book *may be* reproduced for educational purposes. This includes use in courses, journals, websites, and other on-line educational material.

However, the author must be given full credit, with author's name and the title of this book, such as "Model and illustrations created by Mark Fennell, as published in *Mysteries of Electromagnetic Energy: Definitively Solved and Simply Explained.*

When possible, a direct link to the book at Amazon will be provided.

For all other uses, including uses which create a profit or increase the public name for the individual, this author must be contacted: markpoet@aol.com

Preface
Creation and Emission of EM Energy

Overview

This book is the second in a series on my discoveries related to electromagnetic energy, energy fields, and subatomic particles.

The primary focus of *The Creation and Emission of Electromagnetic Energy* is exactly as the title suggests: explaining and illustrating exactly how electromagnetic energy is launched from an electron.

Yet this book is so much more. In the process of explaining how electromagnetic energy is emitted we cover so many more significant discoveries. Among these discoveries you will find:

- The solution for particle-wave duality for the electron.
- A new understanding of electron structure and motion.
- An advanced understanding of electrical current.
- New models of molecular bonds.
- The exact process for photon emission.
- The reason why one photon is emitted rather than another.
- My General Principle of Energy Transfer.

and, of course:

- My General Principle for Particle-Wave Duality.

Each of these discoveries is significant. Together, they provide a rich understanding of energy, electrons, and more.

I have done my very best as a science teacher to make these discoveries as accessible as possible to readers of all backgrounds. Every concept is explained as clearly as possible, with numerous full color illustrations. Key concepts and discoveries are summarized in boxes, as well as a full summary of all points at the end.

A few of the significant discoveries presented in this book will be highlighted briefly below.

Particle-Wave Duality of the Electron: Solved and Demonstrated

One of the most significant discoveries presented in this book is the solution for the particle-wave duality of the electron. For almost 100 years, the electron was considered to be both a particle and a wave. Yet nobody knew exactly why, until now!

This extended into the view that the electron was mostly a wave, which led to further elaborations such as the Schrodinger Wave Equation, the Probability Wave, the Standing Wave, and the Electron Cloud.

For the first time anywhere, the true physical nature of the electron, as both particle and wave, is fully understood and illustrated. I demonstrate for you, step by step, exactly how an electron as particle creates the wave-like patterns.

I explain and illustrate the true physical nature of the electron as wave in an orbital. This leads to a detailed explanation of the physical nature of the associated wave concepts of Standing Wave and the Electron Cloud.

I then show how the electron creates a wave-like pattern as a free electron. (The process is different, and must be distinguished).

Again, for almost a century nobody has been able to accurately explain and illustrate the physical nature of particle-wave duality for the electron. Therefore, this is a significant discovery and presentation to the scientific community.

General Principle of Particle-Wave Duality

This leads to the General Principle of Particle-Wave Duality. I created this principle by combining three individual discoveries:

1. The physical nature particle-wave duality for electromagnetic energy. (First presented to the public in Book 1 of this series, with more detailed explanation in Book 3).
2. The physical nature of particle-wave duality for an electron in an orbit.
3. The physical nature of particle-wave duality for a free electron.

I have combined these discoveries into a single General Principle, which explains all of particle-wave duality.

To put the significance of this General Principle in perspective, remember the following:

For over a century, the particle-wave duality of electromagnetic energy was known, yet none of the scientists could figure out why (including Einstein, Bohr, and Heisenberg). I have solved this mystery.

For almost a century, the particle-wave duality of the electron was known, and the debate on the true nature of the electron has been vigorous. I have figured out the structure of the electron. And I am the first person to demonstrate the actual cause of particle-wave duality of the electron.

And now, I have also put all of these discoveries together into a short General Principle of Particle-Wave Duality. This is presented here for the first time.

General Principle of Energy Transfer

Another important principle I have developed is the General Principle of Energy Transfer.

After spending so many years working with energy, I have come to understand the processes of energy in many arenas. From this, I have developed the General Principle of Energy Transfer.

In one simple sentence, I state exactly what energy transfer is, for the majority of process in the universe. Simple, yet profound. The applications of this Principle will be numerous.

Advanced Understanding of Electrical Current

You may think that electrical current is already well understood, but it is not. I have made several discoveries related to the actual process of electrical current. In this book you will learn many aspects of electrical current which are much more physically accurate than what is being taught today.

Any one of these discoveries would be significant; in this book you have all of them. These many discoveries are presented together, thus giving you at one setting a significantly more accurate understanding of electrical current.

Structure and Motion of the Electron

Several sections of this book discuss the structure and the motions of the electron. Understanding the structure and the motions of the electron will lead us to full understanding of how emission of electromagnetic energy occurs. This will also help us understand the particle-wave duality of the electron. Therefore, understanding the structure and motion of the electron is very important.

Many details will be introduced and illustrated in this book. However, do note that a later book will have full illustrations of the composition of the electron, and full illustrations of all electron motion.

Yet, do know that I have worked out the entire structure of the electron. Know that I have worked out the entire sets of motions of the electron (causes, directions, and strengths).

All of these concepts are new to the world. The most important of these concepts are introduced here for the first time.

Creation and Emission of EM Energy

The primary topic of this book is the creation and emission of electromagnetic energy.

You probably know that an electron releases excess energy by emitting a photon. Yet how *exactly* does the electron emit that photon? For the first time, the process is completely understood.

Therefore, after you understand the structure and components of the electron, after you understand the cause of electron motion, and after you understand the General Principle of Energy Transfer, then you are ready to understand exactly how an electron will emit a burst of electromagnetic energy.

The exact process is understood for the first time, and here fully explained and illustrated.

A similar question is: why will an electron emit one frequency photon over another? This too is understood and explained for the first time. The answer is in a concept I call the "Threshold Percentage", which I discuss frequently throughout this book.

Power Lines and Antennas

Electromagnetic energy is usually emitted by: power lines, antennas, molecular bonds, or atomic electrons. In this book I show the process for each.

I first explain the process of EM emission from a power line. (This is where the discussion of electrical current is important, along with all the new discoveries related to current). I then show how the radio antenna is similar to the power line, and therefore explain the process of radio antenna transmission as an extension of the emissions from a power line.

I also offer a few practical points related to the emission of electromagnetic energy from power lines and antennas.

Molecular Bonds

Molecular bonds are significant sources for emission of electromagnetic energy. Therefore these molecular bonds are discussed and illustrated in detail.

Unique to this book are several new models of molecular bonds. These new models explain simply and accurately the process of molecular vibration. These models also explain the process of electromagnetic energy emission, from any one electron in a molecular bond.

All of these models and explanations are new to the world.

Contribution to the Next Generation

In total, this book has numerous discoveries, many of which will undoubtedly be significant assets for the scientific community. I am pleased to offer this collection of discoveries to fellow scientists, to students, and to all curious readers.

I have also done my very best work as a science teacher to make these discoveries as accessible as possible to readers of all backgrounds. Every concept is explained as clearly as possible, with numerous full color illustrations. Key concepts and discoveries are summarized in boxes. In addition, there is a full summary of all points at the end.

It is my hope that these discoveries become standard concepts for the next generation of thinkers, and that the illustrations become commonplace teaching aids for the future.

May you be as enlightened and excited by reading these discoveries as I have been in providing these discoveries to you.

Mark Fennell

Author's Personal Introduction to the Reader

I have been interested in electromagnetic energy most of my adult life. It is an amazing entity, with numerous properties and the ability to perform a variety of interesting phenomenon.

Furthermore, I have come to realize over the years how important electromagnetic energy is to our lives. Electromagnetic energy is used both in the natural world and in man-made devices (and in thousands of ways). Electromagnetic energy is used to produce light, to create color, and to see into the most remote locations. It is used for art and entertainment, for health and medicine, and for communication. Indeed, I have come to realize that electromagnetic energy has a profound impact on most of our personal experiences.

However, it is only recently that I truly understand the nature of electromagnetic energy. This deep understanding is what I present to the public today.

Certainly I studied electromagnetic energy in college, yet I was never satisfied with their descriptions. To start, most descriptions are abstract, or worse, very highly mathematical. I wanted a more physical description, something that made sense as a physical object. If you could hold a ball of light in your hands, to look at it and study it, this is what you would see.

Thus, the drawings and descriptions in this book represent physical realities, physical objects, which we can comprehend in a very practical way.

Note that scientists such as Bohr felt the same way about certain topics. This is what led such eminent scientists to make advancements in our understanding of science. In the same spirit, my desire for understanding anomalies in the current understanding of electromagnetic energy, led me to make further advancements in the understanding of science related to EM energy.

In fact, I like to think that what Bohr has done in advancing our understanding the atom, I have done for understanding electromagnetic energy.

Regarding the traditional explanations: I understood what the texts were trying to tell me. I understood the equations (as complex as they were), and could use those equations for my own calculations. Yet I wasn't satisfied with the explanations. Something was missing.

Since college I have read about electromagnetic energy in various magazines and books. I have studied it, and I have worked with it over the years.

Yet something was amiss. At the very least was the central debate of "is it a particle or is it a wave?" (I now have a definitive answer to that question, with my own insights on the process).

There were many other questions as well. For example, consider the concepts of penetration and size of the energy. Wavelength cannot correlate with size of the energy, because wavelength is a width, and the penetration size is related to diameter.

These and many other puzzles concerned me over the years. There were also many other anomalies I wanted to resolve. It is only in the last year that I have figured out all of the answers.

At the same time I became very much interested in electrical power. This became my passion. I devoted myself to studying every technology related to electrical power, I talked with experts, and I visited facilities. And in this area too I made some unique discoveries. (These books and discoveries are now published in the series "Energy Technology Explained Simply").

During this detailed study, I learned details about the generation and transmission of electrical power. This includes alternating current and voltage on the atomic scale, magnetic fields, and so much more.

I also started reading the works of Tesla. This includes both his inventions and his general descriptions of phenomena.

Then suddenly in February-March of 2012 everything made sense. Within a few weeks, everything became clear. I devised new theories on the fundamental nature of electromagnetic energy. From these models and mechanisms I was able to explain all other aspects of electromagnetic energy. The new models could explain everything!

A second burst of insight came in October-December 2012. Many more insights came throughout 2013. These were additional insights which incorporated additional models and additional explanations with the previous new models, in order to explain more aspects of observed EM phenomena.

Finally, it all made sense. All characteristics of electromagnetic energy are understood. All physical properties are understood. With just a few breakthroughs in understanding, all properties of electromagnetic energy can be explained.

In your hands you have the new models. In your hands you have an easy-to-understand set of guides on electromagnetic energy.

I provide you with unique analogies which will help you understand all traits of electromagnetic energy. I provide numerous drawings, so that all physical characteristics and all processes are easily understood.

In these series of books you now have a complete description, an accurate understanding, of all characteristics of electromagnetic energy. I hope that this deeper understanding of electromagnetic energy will enrich your life.

M.F.

List of Books for the Series
Mysteries of Electromagnetic Energy: Definitively Solved and Simply Explained

Part 1: Fundamental Properties of Electromagnetic Energy

Part 2: Creation and Emission of Electromagnetic Energy

Part 3: Frequency and Pulsation of Electromagnetic Energy

Part 4: Penetration and Absorption of Electromagnetic Energy

Part 5: Communication Using Electromagnetic Energy

Part 6: Diffraction, Interference, and Red Shift

Part 7: Particle Wave Duality

Part 8: New Models of Electrons, Orbitals, and Atoms

Table of Contents
for
Mysteries of Electromagnetic Energy, Part 2: Creation and Emission of EM Bursts

7. <u>Creating EM Bursts: Threshold % and Driver Strings</u>   17

a. Introduction

b. Electrons and Energy Strings

c. The Energy Percentage

d. The Energy Percentage Threshold or Threshold Percentage

e. Launch Energy and Inherent Energy of the EM Burst

f. Launch Energy and Inherent Energy Related to Mass

g. Two Step Process of Launching EM Burst

h. Detailed Process of Launching EM Burst

i. Motions and Energies of the EM Burst in Space

j. Review of Launching EM Burst

k. Electron Motion versus Energy Fields (Driver and Non-Driver Strings)

l. Energies of the Electron System Revisited

m. Lunching the EM Burst: A more Sophisticated Understanding

8. EM Bursts from Power Lines and Antennas   37

a. Introduction

b. Review of Strings, Fields, Drivers, Motions and Threshold Percentage

c. EM Bursts from Power Lines: in Brief

d. Creating Electrical Current: Details of the Process

e. High Voltages, Driver Strings, and Electrical Current

f. Creating Alternating Electrical Current

g. General Principle of Energy Transfer

h. Electric Fields and Electrical Current

i. Directions of Electrical Fields and Electrical Current

j. Magnetic Fields and Electrical Current

k. Emission Process of EM Burst from Power Lines - in Specific Detail

l. Power Loss from EM Bursts into the Air

m. Radio Transmission Antennas

n. Factors which Affect Frequency Emitted from Transmission Antennas

o. Fraction Antenna Size

p. EM Bursts are Independent After Emitted

q. A Few Points on High Voltage and EM Bursts

9. Creating EM Bursts: Molecular Vibrations　　　　73

a. Introduction

b. Molecular Vibrations Creating EM Bursts: General Overview

c. Current and Fields in Molecular Bond Vibrations

d. Molecular Bonds as Parallel Wires

e. Molecular Bonds as a Racetrack: The Elliptical Looping System

f. Two Molecular Orbital System

g. Process of Molecular Vibrations Creating EM Burst

h. Additional Factors in the Creation of EM Bursts from Molecule

i. Frequency of EM Burst is Related to Speed of Electron

j. Loss of Energy When Burst Emitted is Due to % of Diverted Energy

k. Amount of Time in Higher Energy Before Emit

l. Higher Energy Orbitals: How an Orbital System Exhibits Acquired Energy

m. Schematic of EM Absorption and Subsequent Emission

n. Energy Level Schematic Understood with Energy Strings

o. Energy Percentage Determines Which Option of EM Burst will be Created

p. Starting from a Different Level of Energy

q. Steps of Sequential Bursts and Percent Energies

10. Electrons as Particles and Waves — 113

a. Introduction
b. Two Types of Electrons
c. Atomic Electron as Particle and Wave (Overview)
d. Electron as Particle and Wave on Power Lines
e. Electron as Particle and Wave in Orbitals
f, Electric Fields Always Face the Nucleus
g. Complex Orbital Shapes and Complex Wave Patterns
h. Molecular Orbitals and Fields Pointing Toward Nuclei
i. Standing Waves of Electrons
j. Electrons as Blurred Energies
k. Creating the Electron Wave for a Free Particle
l. Two Methods of Electron as Particle and Wave Reviewed
m. General Principle for Particle-Wave Duality

Chapter Summaries — 143

a. Overview of the Summaries	143
b. Chapter 7 Summaries	145
c. Chapter 8 Summaries	161
d. Chapter 9 Summaries	181
d. Chapter 10 Summaries	205

Chapter 7
Creation of EM Bursts:
Threshold Percentage and Driver Strings

Introduction

Overview

At this time we will discuss the creation of electromagnetic bursts. A key concept for emission of EM Bursts is the "Threshold Percentage". When this Threshold Percentage is reached, then a burst of electromagnetic energy will actually be launched.

The "Threshold Percentage" is a new concept which I developed. In this chapter I introduce and explain this Threshold Percentage in detail. In fact, most of this chapter is written to deliver the Threshold Percentage as the end result.

Basic Mechanism

Most sources for EM bursts create these bursts using the same general mechanism. This mechanism requires:
1. Alternating electrical current, and
2. A high enough energy to meet the threshold for emission.

Concepts Discussed in This Chapter

In this chapter we will look at the most fundamental principles behind the emission of electromagnetic energy. Those principles include: a) driver and non-driver energy strings, b) electron motion versus energy fields, c) the "energy percentage", and the "Threshold Percentage".

We will also discuss different types of energy associated with emission, including the "inherent energy" and the "launch energy".

All concepts discussed here will apply to any emission of electromagnetic energy from any source. This includes emissions from power lines, from transmission antennas, and from molecules.

Electrons and Energy Strings

We will discuss new models for the electron in later chapters. However at this time, in order to understand the creation of EM burst from a power line, atom, or molecule, we will cite some important structural elements of the electron.

The electron is essentially an object, like a planet. Attached to this electron are energy strings. The electron has a variety of motions, including forward motion, spin, and vibration. The energy strings are like Lego blocks: the energy strings can be built up, taken apart, and rearranged.

The most important thing to know at this time is that whenever new energy is added to the electron, some of the energy will go into the arrangement of the energy strings, and the rest of the energy will go into the electron motion.

The Energy Percentage

Added Energy Can Be Diverted to Two Areas

Whenever we add energy to a power line, the energy can go into two areas: 1) the motions of electrons, or 2) the strings of the electrical field.

1. Energy of Electron Motion: Most of the energy in a power line will be diverted to the motion of the electrons. The energy will be put into any one of the motions which the electrons may have. Thus, the electron will vibrate faster, spin faster, or travel forward faster.

2. Strings of the Energy Field: Some of the energy added will be diverted to the strings on the energy field. Thus, as we add energy to the system, some of that energy will go into building more energy strings. Note that the added energy can also create thicker energy strings or longer energy strings, in addition to creating a larger quantity of energy strings.

Therefore, every time we add energy to a system where alternating current flows, the energy can be added to either 1) the motions of electrons, or b) the electric or magnetic energy fields.

We can now define a new concept: The Energy Percentage

> **The Energy Percentage**
> is the amount of energy diverted to the energy fields
> (rather than electron motion)
> as compared to the total energy of the system.

Percentage of Energy Diverted to Each Activity

Now we come to the concept of percentage of energies diverted to each activity.

If we add 100 Joules of energy to our system, we know that some energy will go into the motions of the electrons while the remaining energy goes into the strings of the energy field. The question is what percentage of the original added amount will go into each area?

This is what makes the difference between a power line and a transmission antenna. Technically, the transmission antenna is simply a piece of wire, same as the power line. Also, both wires have alternating current flowing through. So what is the difference? The difference is the percentage of the added energy which is diverted to electron motion versus the energy field.

If most of the added energy is diverted to the motion of electrons, then you will have more of a power line than a transmitter. In this case, the additional energy will provide greater power. [Power is current x voltage. And we said that voltage is the energy of electrons. Therefore, add energy to the electrons, you will have more power in your power line].

At the same time, if most of the additional energy is diverted to the electrons, then very little will be diverted to the energy fields. Therefore, although the power line will continue to create alternating energy fields, these fields will be very small.

The second possibility is to divert most of the additional energy to the strings of the energy field, rather than to the motion of electrons. Now we have a transmitter rather than a power line.

In this case, the alternating energy fields are very strong. This can be due to many energy strings, thicker energy strings, or the energy strings may move further out before reversing back to the wire.

And if enough energy is diverted to the Energy Fields, then this will produce a burst of EM energy.

Additionally, the remaining energy which is left to motions of electrons is very small. This results in very low voltage, very weak vibrational energy, and much slower forward motion.

Therefore, when we add energy to a system of alternating electrical current, some energy will be diverted to the motion of electrons, while some is diverted to the strings of the energy field. The percentage of added energy which is diverted to each area will create either:
1) higher energy electrons (and thus greater electrical power), along with much weaker energy fields, or
2) stronger energy fields, possible emission of EM burst, but electrons with much lower energy.

Note that this situation exists for every object where alternating electrical current flows.

The Energy Percentage Threshold

Energy Percentage Threshold Required for Emission of EM Burst

Notice that as of yet no energy field has left the wire. In all our discussions so far, we have created alternating energy fields, yet at no time has any energy field separated from the line. This is about to change.

Whenever we have alternating current, most of that energy is applied to the energy of electrons (not to the energy fields). Therefore, we create alternating energy fields as we create alternating current, but most of the time these energy fields do not leave the wire. These fields extend out a certain distance, but then come back. These fields are tethered to the wire, like an animal on a rope. They do not leave.

However, if we give our energy fields enough energy…then the energy fields can break loose.

This is very much like an animal tethered to a post. He keeps pulling and pulling…and then finally one day he pulls with enough energy to break the rope from the post. At this point he is free! He starts running, running, running forward.

The same thing happens with our energy fields. They keep stretching forward, but are tethered to the power line. (And in addition, they are being pulled in the opposite direction). Finally at one point the energy field has enough energy to break free!

Like the animal on a rope, the fields stretch to their maximum (as they normally do)…but this time the field has just a bit more energy…just enough to break free from the line. Boom! The energy burst is off. Like the animal running far away, the energy burst is now free and independent. It is not tethered anymore.

Notice that in order for the energy field to break free we need a minimum amount of energy. This is the threshold energy.

More specifically, it is not just a particular amount of energy. Rather it is the % of energy. When additional energy is put into our system, the majority of that energy must be put into the energy field in order for the energy fields to separate from the wire.

Stated another way, the relative percentages of additional energies in the system must be such that more is put into the energy field than into the electron motions.

Only when the amount of energy in the energy field reaches this "threshold percentage" (relative to the total amount of energy in the system) will the energy fields be able to leave their tether, and become completely free.

Only when the % of energy reaches this threshold percentage will the energy fields be able to take flight, and be able to lift-off.

It is only at that point that a burst of EM energy will be created.

> The Energy Percentage Threshold,
> or the Threshold Percentage
> is the required amount of energy to be diverted to the energy fields (rather than electron motion) as compared to the total energy of the system, in order to create a free flowing burst of electromagnetic energy.

Threshold Percentage =

$$\frac{\text{Amount of Energy in Energy Fields Required for Lift-Off}}{\text{Total Energy of the System}}$$

Launch Energy and Inherent Energy of the EM Burst

Launch Energy and Inherent Energy: Overview

When the EM burst is launched, it actually has two energies. It has a forward trajectory energy (like the baseball being thrown). It also has the pulsation energy.

Therefore, the total energy which is applied to the EM burst at the time it takes flight, will be divided into these two areas. I refer to these two energy types as "Launch Energy" and "Inherent Energy".

The "launch energy" is the energy required actually launch the EM burst from the electron into the air. The launch energy creates the energy of forward motion.

The "inherent energy" is the energy which the EM burst has when launched. This Inherent energy is what creates the frequency (rate of pulsation) of the EM burst. The inherent energy is the energy of the EM burst when it is an independent entity, and will remain constant until the EM burst hits an object.

Understanding these energies are important if we want to more fully understand how EM bursts are created.

Energy Types are Akin to Particle Wave Duality

Notice that the Total Energy of the Electromagnetic Burst is a sum of the Kinetic Energy of the Forward Trajectory (like the flight of the baseball) and the Kinetic Energy of Pulsation.

This means that the two energies (launch energy and Inherent energy) are very much akin to the Particle-Wave Duality of the EM burst. There is an energy amount in the particle nature of the EM burst, and there is energy amount in the wave nature of the EM burst.

Both energies must be taken into account if we want to know the total energy of the EM system. And therefore, just as EM is both a particle and a wave, the energies of the EM burst are two energies: one applied to the particle nature (forward motion), and one applied to the wave nature (pulsation).

> The Launch Energy is the amount of energy required to launch the energy strings from the electron.
>
> The Inherent Energy is the amount of energy in the EM burst as an independent entity.

Launch Energy and Inherent Energy Related to Mass

Overview

Both types of energy (inherent energy for frequency of pulsation, and forward trajectory energy from launch) are related to the overall mass of the strings. The explanations and reasoning is slightly different for each, and yet both are ultimately related to the mass of the energy strings.

> Both the Inherent Energy and the Launch Energy
> of the EM Burst are related to
> the Total Mass of the Energy Strings

Both pulsation and forward motion are motions of the EM burst. It is the combination of both motions which create the wave pattern. Therefore we should spend some time looking at the energies involved in both motions.

Launch Energy = Forward Kinetic Energy

The Launch Energy is in fact the kinetic energy of forward travel. That is, the energy which is used to launch the EM burst into the air is then applied to the forward motion of that EM burst.

Forward Trajectory Energy Directly related to Mass

The EM burst can be considered a particle. It flies forward, much like a baseball. Therefore, the forward energy of the EM burst is essentially the same as the forward energy of the baseball thrown across the field.

On a simple level we can see how the launch energy is related to mass. For example, If we launch a rocket into space, the amount of energy required depends on the mass of the rocket. A rocket with greater mass will require more energy to launch. In the same way, an EM burst with greater mass will require more energy to launch.

We can also look at the kinetic energy of the EM burst (as the baseball-like particle). The kinetic energy of an object flying through the air is a combination of mass and velocity. In the simplest form, the calculation for this kinetic energy is: $KE = \frac{1}{2} (mass)(velocity)^2$. Therefore, the kinetic energy of the forward motion of the EM burst is directly related to the mass of the EM burst.

Forward Trajectory Energy ONLY related to Mass

One of the interesting properties of electromagnetic energy is that the forward velocity of all EM bursts are identical. That is, for any particular traveling medium (such as space or pure water) all frequencies of electromagnetic energies travel forward at the same rate.

In other words, we can consider all EM bursts as simple and identical baseballs, when we look at the forward traveling speed. The frequency of pulsation has no effect.

Therefore, since the velocity is all same for every EM burst, it is only the mass which makes the difference in Kinetic Energy. Consequently, EM bursts with greater mass will have greater kinetic energy for the energy of forward travel.

Internal Energy = Pulsation Kinetic Energy

As stated earlier, the Internal Energy is the total energy of the strings in the EM burst. The specific value of the internal energy will result in the specific pulsation frequency. (This will be discussed in detail in later chapters).

Kinetic Energy of Pulsation Related to Mass

Therefore, the pulsation energy is related to both the energy of the strings and to the mass of the strings.

The process will be described in detail in later chapters. For now just know the following: energy strings have mass. More strings in the EM burst will result in greater overall mass. Also, larger strings will result in greater overall mass. Therefore the overall amount of mass of the EM burst is essentially the same as the overall amount of internal energy of the EM burst. Then, this mass will drive the process of pulsation, with greater mass creating a faster pulsation. (Details will be shown and explained in a later chapter).

Thus we can say that the pulsation energy of the EM burst and the overall mass of the EM burst are directly related.

Both Kinetic Energies are Related to Mass of Strings

Thus, both types of kinetic energies (Internal Energy and Launch Energy) are related to the mass of the energy strings. And yet the reasoning is different for each.

1. We begin with the internal energy of the EM burst. This is the group of energy strings that will become an independent entity when launched. They are essentially grouped, ready for launch, but not yet enough energy to launch.

Because energy strings are both energy and mass, greater energy will always have greater mass.

The mass of the internal energy strings is what drives a specific pulsation frequency. Thus, the amount of internal energy means a certain amount of mass, which then means a particular pulsation frequency.

2. The mass of the internal energy strings also determines the required Launch Energy.

The launch energy is the amount of energy required to launch the EM burst from the surface of the electron. Thus, the grouping of energy strings is arranged, just waiting to be launched. All that is needed is the right amount of additional energy to give the strings that extra "push".

The exact amount of additional energy required will depend on the mass of the EM burst to be launched. If the grouping of energy strings has greater mass, then we will require greater energy to launch that group as the EM burst.

Thus both the internal energy and the launch energy relate to the total mass of the energy strings in the photon. The two processes are somewhat independent, and yet both depend on the total mass of the strings in the final EM burst.

Correlations Between Kinetic Energy of Forward and Pulsation

This brings us to some correlations between the kinetic energy of the forward trajectory and the kinetic energy of the pulsation.

1. There IS INDEED a correlation between KE of the forward trajectory, and the KE of pulsation. Both of these kinetic energies are related to the total energy of the energy strings, and to the total mass of those energy strings.

2. This brings us to another correlation as well: the same burst which has higher KE for forward trajectory also has higher KE for pulsation.

The higher kinetic energy for pulsation exists because the energy strings contain much more energy. Furthermore, these higher energy strings have greater mass, which is responsible for the faster pulsation.

The higher kinetic energy for forward motion also exists because of the mass of the strings. Because the energy strings have greater energy, they have greater mass. Then because the launch energy required is dependent on the mass (see above) a greater launch energy is required. The launch energy of course becomes the kinetic energy of the particle in forward motion. Thus, an EM burst with greater mass has much greater kinetic energy for forward motion.

The net result is as follows: Any EM burst with greater internal energy will have both greater kinetic energy of pulsation and greater kinetic energy in its forward motion.

Preview of Mass and Diameter of EM Bursts

The diameter of an EM burst is a new territory of discovery, and will be explained in a later chapter. At this time it is good to give a preview: At this time, note that the EM bursts with highest mass are those which are the SMALLEST diameter, and also have the highest frequencies.

Thus, you can see many correlations between mass of energy strings, energy of pulsation, frequency of pulsation, and energy of forward motion. Again, the details of these concepts will be discussed throughout the book.

> The EM bursts with the greatest mass will have the smallest diameter, and the fastest pulsation frequencies.
>
> The EM bursts with the least mass will have the largest diameter, and the slowest pulsation frequencies.

Two Step Process of Launching EM Burst

Inherent Energy and Launching Energy: Overview

The actual launch of the EM burst is a two-step process: 1) diverting an initial amount of energy to the energy strings, then 2) adding enough energy to the strings so that they can break free from the electron.

Therefore, for an EM burst to be launched, we need both inherent energy and launching energy.

Restated from above, but in a slightly different way: the "inherent energy" is the amount of energy the EM burst has when it is an independent entity in space. This is the energy the EM burst has which determines its pulsation frequency and its color. In contrast, the "launch energy" is the amount of energy required to launch that entity into space.

Note that the "Threshold Energy" is actually a combination of the inherent energy and the launch energy.

> The "Threshold Energy" includes both the Launch Energy and the Inherent Energy.

Building the EM burst before Launch

Before an EM burst can be launched, the energy strings must be given a certain amount of inherent energy. This is essentially building the EM burst. Similar to building the rocket before launch, we are building the EM burst before launch.

And just as we can build different size rockets to send into space, we can create different amounts of size EM bursts before sending into space. Specifically, we are creating EM bursts with various amount of inherent energy. This inherent energy is built into the number and thickness of the energy strings.

Mass of the Inherent Energy

All energy has mass, and all mass has energy. This was shown by Einstein in his famous equation $E = mc^2$. This is also true of energy strings.

Each energy string has a certain amount of mass. And thicker energy strings will consequently have greater mass. Therefore, any energy string which is thicker will in fact have greater energy and greater mass.

And just as we can build rockets of different mass to send into space, we can create different EM strings of different mass to send into space.

This mass, in turn, will dictate how much "launching energy" is required to emit the EM burst.

Detailed Process of Launching the EM Burst into Space

Therefore, again, we first build our potential EM burst (without launching) just as we must build a rocket before we can launch it. Building the potential EM burst prior to launching is essentially creating the thickness and number of energy strings.

These energy strings remain attached to the electron, just as the rocket remains connected to the ground. In order to emit the strings as an EM burst we must apply launch energy, just as we would to launch a rocket from the ground.

The amount of energy to launch the EM burst depends on the mass of that burst. This is the same as the amount of energy to launch a rocket depends on the mass of that rocket.

The basic process of launching an EM burst occurs in the following steps:

a. Energy must first be diverted into the Inherent Energy. This means that the energy is diverted to the energy strings rather than the motion.

b. This will define the frequency of pulsation when the burst is launched.

c. This amount of inherent energy then defines the mass.

d. The amount of Launch Energy required to emit the burst is then based on that mass.

e. When enough energy has been diverted to the energy strings to launch the strings of that mass from the electron, then the EM burst will be emitted.

Motions and Energies of the EM Burst in Space

Overview

The two types of energies (inherent energy and launch energy) will continue to play a role while the EM burst is flying through space.

Remember that the EM Burst has two motions: forward motion and pulsation motion. The forward motion, like a baseball, is determined by the launch energy. The pulsation motion is determined by the inherent energy.

Direction of Travel for EM Burst

Notice that the direction which the energy field is stretching at the time it breaks free…is the direction of forward travel for the energy burst. That direction which the energy strings stretch at the time they manage to break free from the wire…that is the forward direction which the burst travels (like the direction of the baseball thrown).

Review of Launching EM Burst

Thus, we first build our system of energy strings, with a certain number of strings and thickness of strings. This set of energy strings will become the inherent energy for the EM burst once it is launched.

Yet the strings remain attached to the electron. They are tethered to the electron. The strings must have additional energy, beyond the "inherent energy" that will be in the EM burst as independent entity, to actually lunch the EM burst. Enough additional energy must be applied to those strings for those strings to break free from the electron. This additional energy is the launch energy.

The amount of energy required for launch depends on the mass of the EM burst. This mass is the mass of the energy strings. (All energy has mass, and all mass has energy, therefore the thickness and number of energy strings determines not only the amount inherent energy, but also the mass of the EM burst.

The Threshold Energy is the energy required to launch the EM burst into space. We know now this Threshold Energy is a combination of 1) the inherent energy – in the strings when the EM burst is independent entity, and 2) the launch energy - the amount of additional energy to launch the mass of that inherent energy. When this Threshold Energy is reached, the EM burst will in fact be launched.

Electron Motion versus Energy Fields:
Driver Strings versus Non-Driver Strings

Introduction

We have often discussed the concepts of energy being diverted to the energy strings versus the motions of the electron. Now it is time to get a more detailed understanding of what this transfer of energy means.

In brief: most of the motions of the electron are actually driven by energy strings. Thus we have two sets of energy strings: drivers and non-drivers. Energy strings which are drivers will drive many of the motions of the electron. Energy strings which are non-drivers will extend into the air, and be measured as electric and magnetic fields.

Many of these energy strings are free to move about. Therefore, some driver energy strings can become non-drivers. Some non-driver energy strings can become drivers.

It is in this way that the actual transfer or diversion of energy from "motion" to "energy strings" (as described above) actually occurs.

A Glimpse from Future Chapters

We will see in later chapters (on the new models of the electron) that the energy strings are what drives the motions of the electrons. Thus, when we talk of energy is being transferred from the energy strings to the motion of the electron, what is really happening is that energy strings are changing their positions and activities.

Specifically, there are some strings which drive the motions of the electron, and some strings which extend into the air. Therefore, when we have "faster motion" we actually have more of the energy strings applying their energy to the motion of the electron. In addition, this means we have fewer energy strings freely extending into the air.

When we have "slower motion" we actually have fewer of the energy strings driving the electron. This also means we have more of the energy strings freely extending into the air.

Electrons in Orbits, in Electrical Current, and After Absorption

The degree to which the energy strings will transfer the roles from driver to non-driver (and vice versa) will depend on what the electron is doing.

1. Most electrons in a stable orbit will have a particular percentage of energy strings driving the electron, versus energy strings extending into the air. This will generally be the status quo for electrons in an orbital.

2. However, electrons in an electrical current, particularly those in power lines or transmission antennas, will transfer their energy strings from driver to non-driver, and from non-driver to driver. This shifting occurs all the time in electrons which exist as electrical current.

3. Also, any electron which absorbs energy (see the chapter on absorption) is in fact adding energy to the electron system. This new energy can be added to the driver strings, which will increase the motion of the electron, or the non-driver strings, which will increase the measured electric and magnetic fields.

Restating Earlier Concepts in Terms of Driver Energy Strings

When we talked earlier of the energy of the electron system being in two areas (electron motion and in energy strings) what really exists is driver energy strings (which create the motions) and non-driver energy strings (which creates the fields we can measure).

Thus we can restate earlier concepts as follows:

1. The "transfer" of energy from "electron motion to energy strings" is really process of the energy strings changing roles from driver to non-driver.

2. The "total energy of the electron system" is in fact a combination of driver energy strings and non-driver energy strings.

3. The "Energy Percentage" is really the percentage of energy in non-driver energy strings as compared to all the energy strings in the electron system.

Review and Summary

Again, the energy strings in the electron system can be drivers, in which they drive the motions of the electron, or the energy strings can be non-drivers, in which they extend into the air and can be measured as electric and magnetic fields.

In addition, many of these strings are also free to migrate within the electron system. This is the actual process by which energy is transferred from energy strings to the motion of the electron, and vice versa.

Energies of the Electron System Revisited

Overview

At this time it is good to look at all the energies of the electron system, and see how they all interrelate.

Electric Field/Strings and Magnetic Field/Strings

There are two energy fields: electric and magnetic. Each field is in fact a set of energy strings. Therefore we have two sets of energy strings.

The fields we measure are actually the strings extending beyond the material through the air, and ultimately reaching the detector.

Driver Strings versus Non-Driver Strings

As discussed above, the energy strings can play two roles: driver and non-driver. The driver energy strings will produce most of the motions of the electron. With more strings in the driver role we will have an electron with faster motions.

The non-driver energy strings will not be involved with the motions of the electron; instead, the non-driver strings will extend into the air and be measured as electric field or magnetic field.

Inherent Energy and Launch Energy

Then we get to the concepts of Inherent Energy and Launch Energy. The inherent energy is the energy contained in the energy strings, when the EM burst is emitted as an independent entity.

However, before the EM burst becomes an independent entity, the burst must be launched. This means there must be sufficient Launch Energy diverted to the energy strings so that they break free from the electron.

And yet these two types of energies are related. The launch energy is based on the mass of the strings, which in turn is related to the total energy of the strings. Thus, the inherent energy of the strings, at any moment, is what will determine the required launch energy to emit the EM burst.

Inherent Energy can Change before Launch

The necessary values will also change until the actual launch occurs. Remember that the inherent energy of the strings is the energy that will be contained in the EM burst, and yet the EM burst is not launched yet. This means that as energy shifts from electron motions to energy strings, and from energy strings to electron motions, the exact amount of energy in the strings will change. It is a highly dynamic situation.

The number and thickness of energy strings is changing from moment to moment. Consequently, the mass of the strings will change moment to moment. And the amount of launch energy will also change from moment to moment.

Imagine if the launch energy were being determined and administered by a person. He looks at the energy strings and calculates for launch. First 5 Joules is required. No, 10 Joules. Wait, 15 Joules. Okay, back down to 10 Joules… and so on.

Thus, in order to launch a burst, the "person" will have to administer enough Launch Energy, for a particular mass of energy strings, at the exact moment the Launch Energy correlates exactly to the mass of the strings. This is not an easy task, because the target value is always moving.

However, when energy strings have enough added energy (launch energy) to allow a particular mass of strings to break free, then that mass of energy strings is actually separated from the electron. The EM burst is launched.

Adding Launch Energy

Energy is added to the existing strings to give them more pull strength. When given enough pull strength, the energy strings will launch. However, as stated above, the energies of the strings change all the time. Therefore the required amount of energy must be given at once, like an injection, for the EM burst to launch.

Otherwise (added in bits not enough to launch), any added energy will simply add to the overall energy of the strings. This is a great way to increase the frequency of pulsation, or to increase the amplitude (as discussed elsewhere in the book). However, this is not the way to actually launch the EM burst from the electron.

Threshold Energy Revisited

Now we return to the Threshold Energy. The "Threshold Energy" is the amount of energy required to launch a particular size EM burst from the electron.

We can now see that the Total Threshold Energy is a combination of the following:
1. Inherent Energy of Electrical Energy Strings
2. Inherent Energy of Magnetic Energy Strings

3. The Launch Energy required to Launch a particular mass of electrical strings and magnetic strings.

Energy Percentage and EM Frequency Launched

We also know that this is a Percentage. Specifically, the Threshold Energy is the Percentage of Energy in the energy strings, required for inherent energy and correlated launch energy…as compared to the total energy of the system. (The other energies are in the motions of the electrons).

Furthermore, when a particular electron is known to emit one of several frequencies of EM bursts, the actual frequency emitted will depend on the particular Energy Percentage at that moment in time.

Summary of Energies

Therefore the total energy system of the electron includes: forward motion, spin, and vibration, each of which cause or are affected by magnetic energy strings and electrical energy strings. All of these energies are interrelated, and the amount of energy diverted to each area changes constantly. However, there is a total value for all the energy in this system, and this will not change until an EM burst is absorbed or launched.

In order to launch a burst of EM energy, first some energy must be diverted to the energy strings (both electric and magnetic). Then the proper amount of threshold energy must be injected into these strings at one time for the strings to break free.

The total amount of energy to break free is the Threshold Energy, and includes both the inherent energy of the strings and the correlated launch energy. The specific frequency of EM burst launched depends on the specific Energy Percentage: Energy diverted to the strings as compared to the total energy of the electron system.

Lunching the EM Burst:
A more Sophisticated Understanding

Now we can understand the process of launching an EM burst on a much more sophisticated level. An EM Burst will actually be launched when the following process occurs:

1. Non-driver energy strings (both magnetic and electric) arrange themselves in a particular way. This becomes the "inherent energy".

2. Driver energy strings migrate to the non-driver energy strings. (This is the transfer of energy from the electron motion to the energy of the main electron strings).

3. In order for the grouping of main energy strings to be launched, they need enough additional energy to break free. This is the Launch Energy.
   This launch energy comes first externally, then internally:

   a. First the electron gains additional energy strings, from an external source.

   b. These energy strings usually take on the role of Driver Energy strings first. Thus, the electron will speed up, vibrate faster, change directions, or change location.

   c. After a while, the energy strings migrate internally. Many of the new driver energy strings gradually become non-driver energy strings.

   d. Thus, the externally added energy is finally shifted from the motions of the electron to the set of main electron strings.

   e. Now the main energy strings (the non-driver strings) have enough energy to break free from the electron.

4. It is at this point that the burst of electromagnetic energy is actually launched. A photon is emitted, and the electron's motion is observed to be less energetic.

Chapter 8
EM Bursts from Power Lines and Transmission Antennas

Introduction

Overview

In this chapter we will look at the creation of electromagnetic energy bursts from power lines and transmitting antennas.

We will discuss alternating current and alternating electrical fields. We will discuss electron motion in relation to magnetic fields. We will explain in detail how electric current is actually created.

Then we will take all these concepts, combined with the Threshold Percentage we discussed earlier, and show how electromagnetic bursts are emitted from a power line.

EM from Power Lines Explains EM Bursts from Antennas

Note that when we understand the process of emitting bursts of electromagnetic energy from a power line, we can easily understand the emission of EM bursts from any other source.

The process for emission is essentially the same for any source, including power lines, transmission antennas, and molecules. In fact the transmission antenna is almost identical to the power line, with just a shift in the Energy Percentage.

Therefore, after we go through the entire process for EM emission from a power line, we will discuss the basic processes for EM burst from a transmission antenna.

EM from Power Lines Explains EM Bursts from Molecular Vibration

Understanding the processes of EM burst emission from a power line can also help us understand the process of EM burst from molecular vibrations.

The molecular bond can considered (at its most basic structure) as a power line. And the electron traveling through its complex orbital path is very much like an electron traveling back and forth in an alternating current. Therefore the emission of EM bursts from molecular vibration is essentially the same as the process in the power line.

Therefore, in the next chapter we will discuss and demonstrate the exact process of emission from molecular vibration. This discussion will be built from concepts presented here.

Basic Mechanism

Most sources for EM bursts create these bursts using the same general mechanism. This mechanism requires:
1. Alternating electrical current, and
2. A high enough energy to meet the threshold for emission.

We will discuss the processes in this chapter.

Review of Strings, Fields, Drivers, Motions and Threshold Percentage

Overview

Before we delve into how the EM burst is created and launched from the power line, we should review some important concepts of the previous chapter.

In particular, it is important to review the following: Energy Strings, Energy Fields, Driver Strings versus Non-Driver Strings, Fields versus Motions, Energy Percentage, and Threshold Percentage.

Energy Strings, Driver versus Non-Driver

An electron is composed primarily of energy strings. Most of these strings are attached to the electron.

There are two types of energy strings: magnetic energy strings and electric magnetic energy strings. The magnetic energy strings are the ones primarily responsible for the motion of the electron.

Energy strings can also play one of two roles: driver or non-driver. The driver energy strings create the motions of the electrons (diagrams are shown in a later chapter). The non-driver energy strings are the ones which extend into the air and can be measured by a device as a "field".

Fields as Non-Driver Strings

I want to emphasize that what we know of as "fields" are actually energy strings which extend beyond the electron. Thus, the magnetic field is actually a set of magnetic energy strings which extend beyond the electron, and to the detector. Similarly, the electric field is actually a set of electric energy strings which extend beyond the electron, and to the detector.

Motions of electrons from Driver Strings

I also want to emphasize that the motions of the electrons are created by the driver energy strings. The details will be presented and illustrated in a later chapter. However, it is important to know at this time that there are some energy strings which push on the interior of the electron system, causing it to move. This movement can be many directions, which results in spin, vibration, and forward travel.

Shifting Roles of Energy Strings

It is also important to remember that the energy strings can shift roles. Thus, some driver energy strings can become non-driver strings, in which case the motions of the electron slow down. Some non-driver energy strings can also become driver strings, in which case the motions of the electron will increase.

The energy strings can also disassemble and reassemble elsewhere, much like Lego blocks.

Therefore the energy strings are constantly moving about, rearranging themselves, and changing roles from driver to non-driver. It is a very dynamic situation.

Total Energy and Energy Percentage

1. Total Energy:

The total amount of energy in all of the energy strings is the total amount of energy in the electron system. This total energy will not change (at least until an EM burst is launched).

2. Energy Percentage:

At any given moment some energy strings will be drivers, causing the motions of the electron, while the remaining strings will be the non-drivers which creates the measurable fields.

The amount of energy, at that moment, diverted to the non-driver role (the fields) as compared to the total energy in the electron is what we call the "Energy Percentage".

However, remember that the movement of these strings within the electron system is very dynamic, therefore the exact value of the energy percentage will change moment by moment.

Inherent Energy, Launch Energy and Threshold Percentage

1. Inherent Energy:

The "Inherent Energy" is the amount of energy in the fields (the non-driver energy strings) of the energy strings that will become independent when launched. This is similar to the size of the rocket before launch.

2. Launch Energy:

The group of strings with "inherent energy" will not become independent until those strings have enough extra energy to break free from the electron. The amount of additional energy required for that group of non-driver energy strings (the fields) to break free is the "Launch Energy"

3. Threshold Percentage and Actual Launch:

When enough energy is diverted from the motion of the electron to the fields, then an EM burst will be launched.

Specifically, launching of the EM burst occurs when the following process occurs:

a. Creating the group of energy strings: enough energy strings are diverted from the motion of the electron (driver role) to the fields (non-driver) to make the inherent energy.

b. Launching the group of energy strings: enough additional energy strings are diverted from the motion of the electron to the fields such that the group of strings can actually break free from the electron.

The total energy required to create and launch a particular group of EM bursts is measured as a percentage: the Threshold Percentage. Thus, the "Threshold Percentage" is the total energy in the non-driver energy strings (the fields) as compared to the total energy in the electron system, such that a particular group of energy strings will be launched.

Note that it is possible for different threshold percentages to be created in the same electron. This is what will produce different frequency EM bursts from the same electron system.

Ready for the Next Sections

Now that we fully understand the concepts of 1) electron motions vs. energy fields, 2) driver vs. non-driver energy strings, and 3) threshold percentage, we can now understand how bursts of electromagnetic energy can be emitted from a power line.

As we come to understand the process of emission of EM bursts from the power line, we will be able to understand the emission of EM bursts from any source in the universe.

EM Bursts from Power Lines: in Brief

Overview

At this time I will provide the description of the process of the EM burst being emitted from a power line.

Note however that this will be a concise version. This version is designed to give you a quick understanding of the process. Further details will be described in sections below.

I also want to remind you that the basic concepts provided here will apply to transmitting antennas and to molecular vibrations (to a certain extent).

EM Bursts Emitted from a Power Line: Brief Description

All power lines are capable of emitting bursts of electromagnetic energy, and power lines which carry high voltage are more likely to emit electromagnetic energy bursts. The basic process is as follows:

1. External energy is applied to the electrons. This energy is applied in the form of magnetic energy strings delivered from a magnet in the generator.

2. The additional energy strings cause the electron to vibrate faster and to move forward, thus creating electrical current.

3. The magnet in the generator rotates the opposite direction, which draws the electron to travel in the reverse direction. Thus, an electrical current is created in the opposite direction.

4. This process repeats, quickly, for as long as there is a power source (coal, hydro, etc.) connected to the generator. Thus, alternating current is created and delivered through the power line.

5. Alternating electrical fields are also created with the electrical current. While the electron moves forward, the energy strings extend upward, and while the electron moves backward, the energy string extend downward. Thus an alternating electrical field is also created along with the alternating electrical current.

6. These electrical energy strings will generally remain attached to the electron, unless given enough additional energy to break free.

7. At the same time, there are driver energy strings (which create the motion of the electron) which become non-driver energy strings (which we register as the field).

8. Some of these non-driver magnetic strings can convert into non-driver *electrical* strings.

9. Thus, driver energy strings gradually become non-driver energy strings - first as magnetic only, then some converted to electric, thereby making more of both electric and magnetic non-driver energy strings. Therefore, both the electric field and the magnetic field of the electron increases.

10. When enough energy strings have been diverted from the motion of the electron (driver strings) to the fields (non-driver strings), and when enough of those non-driver magnetic energy strings convert to non-driver electrical strings, then the entire group of non-driver energy strings will lift off. Thus, a burst of EM energy is created and launched.

11. If the voltage is very high then an EM burst is more likely. This is because there is more energy applied to the electron, and therefore it is more likely for enough energy to be diverted to the group of field strings for launch.

Creating Electrical Current:
Details of the Process

Overview

Note that you can find a detailed explanation of the generation of electrical power in my book "Energy Technologies Explained Simply, Volume 1: Introduction to Electrical Power". Here I will provide a brief yet specific summary, as will be relevant to all future discussions on the emission of electromagnetic energy bursts.

Traditional Understanding of Electrical Current

We will first provide an introduction to electrical current by describing electrical current as it is traditionally understood. The traditional understanding of the process is as follows:

A magnet exists in a generator. This strong magnet produces a magnetic field, which is strong enough to push the electrons in a nearby wire. The electrons, having been pushed forward, thereby create an electrical current.

An alternating current is created when the magnet is rotated 180 degrees. When the magnet is rotated, this magnet pulls on the electrons in the wire. Thus the electrons flow back to the generator, making a reverse current.

Because the magnet is continuously rotating, the electrical current is created in each direction, repeatedly. Thus, alternating current is produced in the power line.

A More Accurate Understanding of Electrical Current

We can now add several new discoveries regarding the creation of electrical current. You will now understand electrical current much more accurately, and in greater detail. The details here are important for all future discussions regarding electrical current and emission of EM bursts.

Thus, a more detailed and more accurate understanding of the generation of electrical power on a power line occurs as follows:

1. We begin with a wire made of copper or gold. The particular metal is chosen because the electrons on the outer orbits of the copper atoms are very loosely held by the atoms. This means the electrons can be encouraged to move easily.

2. A strong magnet is set up to rotate near this wire. The magnetic field extends from the magnet, to the nearest electrons on the wire. This provides an external kick to the electrons.

3. More specifically, some of the magnetic strings leave the magnet and enter the electron system. Most of these energy strings go in the role of driver strings. Thus, those new magnetic strings actually drive the electron forward.

4. Because these particular electrons are loosely held by the atom, it doesn't take much energy to move the electrons from atom to atom. Therefore, the electron moves straight ahead…like a car being driven forward.

5. Notice however that this electron does not travel all the way. Instead, this electron bumps into the next electron, which sends that electron further down the line.

6. More specifically, some of the energy strings of electron 1 will combine with energy strings of electron 2. Thus, energy strings have left the first electron, and joined the second electron. This of course means that the second electron has more energy, while the first electron has less.

7. As before, the energy strings of our second electron will go directly into the role of driver energy strings. Thus, these driver energy strings will drive our electron forward.

8. The second electron will hit its next neighboring electron, at which point more energy strings are transferred…and the process continues.

9. Thus, what we see as an electrical current is in fact a series of electrons bumping. Furthermore, the actual process of electrons bumping is in fact energy strings leaving the first electron to join the second electron. And every time energy strings join a new electron, those strings go into the role of driver electrons, at which point they drive the electron forward.
    Also, because of the particular type of atoms in our wire, the electrons on the outer orbitals are loosely held, thus making it easy for electrons to move forward in a single direction.
    This is how electrical current is created.

A More Accurate Understanding of Electrical Current in a Wire:

1. In a wire, electrons do not actually flow. Rather, the electrons bump into each other, then pass on their energy.

   This is much like a series of dominoes bumping, where we observe the "flow" of dominoes falling, though the dominoes themselves move very little.

2. More specifically, electrical current in a wire is the transfer of energy strings, primarily magnetic energy strings, from one electron to another.

3. As the new electron gains magnetic energy strings, the magnetic energy is observed in two ways:

   a. Some magnetic strings become driver strings. These strings cause the electron to move forward at a fast rate.

   b. Some magnetic strings become field strings. These strings are observed as a stronger magnetic field.

4. The transfer of energy strings from electrons to electrons pushes electrons forward, sequentially. This is what we know of as "electrical current".

5. Electron in a wire are able to do this because the outer electrons are loosely held by the nucleus, thus making it easy for electrons to move forward in a single direction.

   If the outer electrons were more tightly held, then the electrons would simply absorb the energy strings, and likely emit an EM burst, rather than move forward as an electrical current.

High Voltages, Driver Strings, and Electrical Current

Voltage Overview

First let us begin with the defining voltage. A good definition of voltage is "the overall energy of all electrons at a particular location". (This definition is from my book *Introduction to Electrical Power*).

Remember that every electron has many motions, including forward movement, vibration, and spin. We also now know that all of these motions are caused by the driver energy strings. Therefore the total energy of these energy strings is what causes the total energy behind these motions.

This also means that the total energy of the driver energy strings is in fact the voltage.

Thus: the total energy of the driver energy strings *is* the voltage, and this total energy of the driver energy strings is what creates the motions of the electrons.

Amperage Overview

Amperage is a measurement of the number of electrons moving forward per second. However, we can now understand this more accurately.

Remember that electrical current is a series of electrons bumping. Or more accurately, it is a series of transfers of magnetic strings from one electron to another.

Therefore, the number of energy string transfers will give us the number of electrons flowing at the same time. Thus, "amperage" is the number of electrons bumping, or the number of energy strings transfers, per second, which then results in the number of new electrons moving forward per second.

A More Accurate Understanding
Amperage and Voltage:

1. Amperage is the number of electrons bumping and transferring energy strings, per second.

2. Voltage is the overall energy of the electron.
   Specifically this means the total strength of the driver magnetic strings and the driver electric strings in the electron.

Transferring Energy Strings and Energy of Electrons

10. When energy strings are transferred from electron to electron, only a few energy strings are transferred at a time. This produces some observable results.

Remember that additional energy strings have been added to the electron system. Therefore, these additional energy strings are the ones which will be transferred from electron to electron.

However, not all of the additional energy strings will be transferred. For example, 70%-90% of any set of energy strings will be transferred from one electron to the next. Therefore, the next electron will not travel as fast as the previous electron.

Then, as we go down the line, there are fewer additional energy strings being transferred from one electron to the next. For example, if we say that the amount of original energy strings transferred from the magnet to the first electron is 100%, after a few miles of transferring energy strings down the power line, we see that the electrons are receiving only 5% of the original amount. Eventually there will not be enough additional energy strings to pass on, and therefore the electrical current stops.

Using High Voltages for Longer Distances

11. It is for this reason that we use extremely high voltages to send power over long distances.

If we begin with an enormous amount of energy strings in our electrons, then there will still be many energy strings which will be transferred to the next electron, even hundreds of miles down the line.

Creating Alternating Electrical Current:
Details of the Process

12. A reverse current is then created by the magnet at the generator. The energy strings at the generator change direction. Because these energy strings are very strong, this action will pull on any nearby energy strings.

Specifically this means pulling on the energy strings in the nearby electrons. Unlike the strings in the magnet, these strings in the electron are too strongly held to leave. The net result is like pulling a toy on a rope: as you pull the rope, the toy comes with you. And so it is with the magnet and the nearby electron: as the strong magnet pulls on the energy strings attached to the electron, the electron is pulled toward the magnet. This causes the nearest electron to move back toward the generator.

13. Now that one electron has moved, there is a "hole". The nearest atom over would like to have an electron in an outer orbital above that atom. Therefore, an electron moves backward to fill its place.

And again, we have another hole. Thus another electron goes backward to fill that electron's place.

The net result is that all electrons are put back in place exactly as they were before.

14. What of the original energy strings from the magnet? They are now spread throughout many miles of electrons. Each electron has a bit more energy than when it started, but not much, because most of that additional energy was transferred to the next electron, leaving only a few additional energy strings in this electron.

15. While this is going on, the magnet itself obtains MORE energy strings. These energy strings are obtained from the rotating axis which spins the magnet. The rotating axis obtained its energy strings from a set of turbine blades, which obtained its energy strings from a power source, such as the water of hydropower.

Thus, our original power source (such as flowing water in hydro power or flowing wind in wind power) is actually providing new energy strings. These energy strings are transferred from place to place, from electron to electron, and from molecule to molecule, ultimately to the generator in our magnet, and then to the electrons in our wire.

Because of this process, a power line will never run out of energy strings. We can continue to "drive" the electrons forward.

16. At this point we are set to go again: Our magnet has been reset with a new filling of energy strings. Our electrons are in their original positions. We can therefore start again with another push of the magnetic field on the electrons.

This of course means the transfer of energy strings from original magnet to electron 1. These energy strings drive the electron forward, until hitting electron 2, which really means transferring some energy strings to electron 2, and driving that electron forward. Thus, a forward current is again created.

17. Taking all these processes together is how we a) create an electrical current going forward, b) create an electrical current going backward, and c) repeating the process again.

General Principle of Energy Transfer

As a corollary to many of the above statements, I can now present my General Principle of Energy Transfer. This principle states that the transfer of energy for many situations is in fact a transfer of energy strings from electron to electron, or from atom to atom.

> General Principle of Energy Transfer:
>
> The transfer of energy in many situations is actually a transfer of energy strings from electron to electron, or from atom to atom

Electric Fields and Electrical Current

Introduction

We will begin by looking at a power line. Understanding the process of energy fields, electrical current, and the creation of EM bursts from a power line will help us understand all other sources of EM bursts. (This includes EM bursts from transmission antennas and from vibrating molecules).

Electrical Current and Electrical Fields

The purpose of a power line is to transmit electrical current. However, electrical fields are created as well.

To be more accurate: the electric field already exists. It is the strength of the field, and the direction of the field which is related to the electrical current.

As discussed earlier, an "electrical energy field" is actually a set of non-driver electrical energy strings. These strings exist all the time within the electron system.

Recall that for an electrical current to be created, the electron will first absorb additional energy strings. Some of these energy strings will become driver energy strings, and thus propel the electron forward. This creates the electrical current.

Yet at the same time some of those additional strings will become non-driver electrical strings. Thus, the total amount of non-driver electrical strings (and hence the total size of the electrical "field") will be greater.

Thus, we observe electric fields with the creation of electrical current precisely because of the additional energy strings. Some additional energy strings drive the electron forward, while other additional energy strings create the stronger field.

Furthermore, a stronger electrical field is associated with a faster moving electron. Again this can be explained by the additional energy strings. When we add a large amount of energy strings to the electron system, then we will have many additional driver energy strings (which thus makes a faster electrical current), as well as many additional non-driver energy strings (which makes a stronger electrical field).

Directions of Electrical Field and Electrical Current

Overview

As stated above, whenever we create an electrical current have two interrelated processes: 1) the electron moves forward, and 2) the electrical field extends into the air.

Remember that both of these are related to the amount of additional energy strings put into the electron system. Specifically, the amount of energy strings determines both the speed of the electron and the field we measure.

In addition, the flow of the energy strings determines the directions – of both the electrical current and the electric field. Regarding the electrical field: these strings will expand, contract, and change direction as the electron moves. Therefore, it is the strength of the field and the direction of the field which is related to the electrical current.

This will be illustrated below.

Simpler Drawings to Show Fields Related to Current

Note that for simplicity going forward, we will often use simple arrows in the illustrations. We will talk less of the energy strings themselves. Using these simple arrows we can show the direction and strength of the field, especially as related to the direction of the electron moving forward.

Electric Field is Perpendicular to Electron Motion

The electrical field always extends perpendicular to the forward motion of the electron. This is the case in a power line, transmission antenna, or molecular orbital.

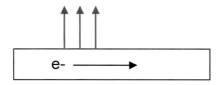

Alternating Electric Fields

The alternating current creates the alternating electrical energy fields. Specifically, when we send electrons in the opposite direction, the energy field will also be created in the opposite direction.

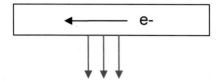

Therefore, when we create alternating current we are also creating alternating energy fields.

For example, when the current goes right, the energy field goes up. And when the current goes left, the energy field goes down. This is how alternating current creates alternating energy fields.

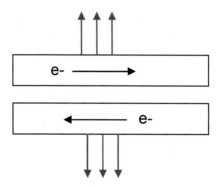

Magnetic Fields and Electrical Current

Magnetic Field as Magnetic Strings

Magnetic Fields are also observed in a power line. As with the electric field, the magnetic field is actually a set of magnetic energy strings.

Magnetic Field Perpendicular to Current and Electrical Field

The magnetic field is perpendicular to the direction of current, as well as being perpendicular to the electrical field. Also, similar to the electrical field, the direction of the magnetic field depends on the direction of current. Thus the magnetic field can extend outward in two directions, depending on the direction in which the electron travels.

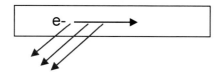

Magnetic Fields, Magnetic Strings, and Motions of Electrons

It is commonly said that magnetic fields are produced by the spin and general vibration of the electron, however throughout this book you will see that magnetic energy strings are responsible for spin.

As discussed throughout this book, energy strings (primarily magnetic strings) can play two roles: driver energy strings (which create the motion of the electron) and non-driver strings (which are observed and measured as fields).

Thus, as one set of magnetic strings drives the spin and forward motion of the electron, the other set of magnetic strings extends outward as the magnetic field.

Emission Process of EM Burst from Power Lines:
in Specific Detail

Introduction
Now that we have discussed additional processes of the power line, including the creating of electrical current and the creation of alternating electrical fields, we can delve into the specific processes involved when an EM burst is created and launched from a power line.

Applying External Energy via Magnetic Energy Strings

1. External energy is applied to the electrons in the power line.
   This external energy is applied from a rotating magnet in the generator. More specifically, some of the magnetic energy strings leave the magnet and enter the electron.

2. Most of the new energy goes into the motion of the electron.
   Specifically, the new additional energy strings become driver strings, which will then increase the vibration and forward motion of the electron.

Creation of Electrical Current

3. If the electron is loosely held by the atom, then a current is created.
   In a good conductor, the outer electrons are loosely held by the atom. Therefore the increase in forward motion actually drives the electron further down the line, thus creating an electrical current.

4. The actual current is electrons "bumping".
   The original electron only goes so far. The actual process involves one electron bumping into the next one, and that electron bumping into the third, and so on, much like series of falling dominoes.

5. **More specifically,** electrical current is transferring magnetic strings, which causes the next electron to move forward.
>   The actual "bump" of one electron to the next is actually magnetic strings being transferred from one electron to the next.
>
>   In this process, the second electron gains more magnetic energy strings. These become driver strings, which causes the motions of the electron to increase – including the forward motion. Thus that second electron moves forward.
>
>   Conversely, the first electron lost magnetic energy strings, which means fewer energy strings as driver strings, and the electron slows down.
>
>   (In a more precise description: some of the field energy strings of electron 1 joined the field energy strings of electron 2. The additional field energy strings of electron 2 became driver strings, and thus propelled the electron forward. And because electron 1 lost some energy strings in its field, then some driver strings became non-driver strings to balance out, which also makes the electron 1 slow down).

Creating Alternating Current

6. The magnet in the generator creates the alternating current.
>   The magnet in the generator rotates opposite direction, which pulls electrons in the opposite direction. This creates a flow of electrons back toward the generator. Therefore, an electrical current is created in the opposite direction.

7. More specifically, the magnetic strings in the generator magnet are pulling the electron backward.
>   The magnetic in the generator rotates 180 degrees. This means that the energy strings in the magnet are flowing the opposite direction. Magnetic strings have a natural pull force for other magnetic strings. Therefore, any nearby magnetic strings will be pulled toward the magnet.

The nearest magnetic energy strings are those strings attached to the nearest electron. And the strings that are "felt" are the non-driver strings (the magnetic field) of the electron. Thus, the strong magnet of the generator pulls on the magnetic energy strings of the nearest electrons.

And, because these magnetic strings are attached to the electron, the electron naturally comes along. Just like pulling the handle of a wagon will also pull the wagon, pulling on the energy field of the electron will pull the electron. The net result is that the electron is pulled backward toward the generator.

8. Reverse current is filling the holes above the atoms.
The sea of electrons in a metal is essentially loosely held electrons above each atom. Note that although each electron is loosely held, an atom still wants an electron there. Thus, if an electron is pulled toward the generator, a "hole" is created above the atom. Therefore the atom will naturally draw the nearest electron to fill its hole. This means that the electron down the way will be pulled back.

Now of course that atom is missing its electron, and pulls on the one next to it. And so on….

It is in this way that a reverse electrical current is created.

9. In total, alternating current is created from magnetic pull and electrons filling vacancies.
The reverse current begins when the strong magnet at the generator pulls on the nearest electrons. Note that "nearest" is relative to the strength of the magnet. A very strong magnet can pull electrons several inches or several feet down the wire. Thus, many electrons can be drawn backward by the magnetic pull alone.

This leaves several vacancies of spaces above atoms. The atoms pull on nearby electrons, which thus is a second process of reverse electron flow.

8. The magnet gets replenished with magnetic strings coming from a power source. This allows alternating current to be created essentially forever.

> The magnet in the generator is continuously replenished with magnetic energy strings. These strings come from the power source.
>
> The power source, such hydro or coal power, will provide energy. This energy is transferred through flowing water or steam, to rotating the turbine and axel, and finally to the magnet itself.
>
> More specifically, it is energy strings which are being transferred from the water or steam, through the turbine, and to the magnet. Thus, the magnet itself is constantly being replenished with new magnetic energy strings. This allows the process to continue, essentially forever.

Magnetic Fields and Alternating Current

9. Alternating magnetic fields are observed with the electrical current. Remember that the magnetic field is actually the magnetic energy strings which extend into the air. When the electron travels forward, the magnetic field will extend in a perpendicular direction, such as outward toward the viewer. Conversely, when the electron travels in the reverse direction, the magnetic field will extend in the opposite perpendicular direction, such as flowing away from the viewer.

10. The correlation between forward direction of the electron and the flow of the magnetic field is related to the general direction of flow for all magnetic strings (driver and non-driver). Some of these points will be illustrated in later chapters. For now note these main points:
    a. Forward direction of the electron is driven by the forward flow of the driver energy strings.
    b. The perpendicular direction of magnetic field is related to the flow of the driver energy strings.
    c. The reverse direction of the electron is created mostly by the pull of the magnet in the generator, or the pull of the nearby atom.
    d. This pull will also change the direction of the magnetic energy strings. This in turn creates a reverse direction of magnetic field and reverse motions of the electron.

Alternating Electrical Fields

11. Alternating electrical fields are also observed with the electrical current.

    Remember that the electrical field is actually those electrical energy strings which extend into the air.

    While the electron moves forward, the energy strings extend perpendicular to the current, such as upward. Conversely, while the electron moves backward, the energy strings extend perpendicular in the opposite direction, such as downward.

12. The strength of the electrical energy field is related to the forward speed of the electron.

    As the electron travels faster, the electrical field extends further outward. As the electron travels slower, the electrical field retreats toward the electron.

13. The electrical field motion expands and retreats and expands again according to electron speed and direction.

    Thus, for example: as the electron travels to the right, the electric field extends upward. As the electron travels faster its energy strings will extend further into the air.

    Then as the reverse magnetic field is applied, the electron slows down. At this point the electric field retreats inward. As the electron travels the opposite direction, the electrical field will extend into the air, yet downward. And as the electron travels faster, the electrical field will extend further in a downward direction.

    This pattern repeats as long as the alternating current is produced.

14. Cyclic energy field produces EM cycles.

    Note that the cyclic nature of the electrical field will create many results, including having an influence on the frequency of EM burst being emitted.

Energy Transfer, Energy Percentage, and Launch

15. Electrons in a power line will generally not produce EM burst.
    At this point we have electrons traveling back and forth, and we have energy fields extending in alternating directions. However, this situation alone is not enough to produce a burst of electromagnetic energy. These electrical energy strings will generally remain attached to the electron, unless given enough additional energy to break free.

16. Diverting energy strings can create burst.
    Remember that we have driver energy strings and non-driver energy strings. When enough driver energy strings change roles to non-driver energy strings, then an EM burst can be launched.

    Energy strings are always moving about, from driver to non-driver, from non-driver to driver, as well as separating, migrating, and rearranging. Therefore it is possible at any time for enough energy from the driver strings (motion of the electron) to be diverted to the non-driver strings (the fields) in order to create and launch a burst of electromagnetic energy.

17. Magnetic energy strings can convert to electric energy strings.
    Remember that we have two types of non-driver energy strings: electric and magnetic. Both will be part of our "inherent energy" (the grouping of strings that will be in our burst when launched).

    In addition, both sets of strings require enough additional energy ("launch energy") so that these energy strings can break free from the electron.

    It is easy to see how driver magnetic strings can become non-driver magnetic strings, thus providing the extra energy required for launching those strings.

    However, the group of electrical energy strings also requires additional launch energy. At first this may not seem obvious, but the answer is that some of the magnetic energy strings convert into electrical energy strings.

Thus, we start by having several magnetic driver strings change roles to non-driver strings. At this point some of those strings join with the existing magnetic energy strings. This will essentially produce the launch energy for those magnetic strings.

Then the other magnetic non-driver energy strings will convert into electrical energy strings. At this point these new electrical energy strings will join the existing electrical energy strings. This will essentially produce the launch energy for those electrical strings.

18. Strings can launch independently or together.
Note that energy strings can launch independently or as a group. For example, if a few magnetic energy strings acquire enough energy to break free, those string can indeed break free and become free strings floating in the air around the electron. (What happens next is beyond the scope of this book).

However, we are most interested in the launch of electromagnetic energy. This requires a grouping of many electrical energy strings and many magnetic energy strings to be grouped and launched at the same time.

19. The actual launch of the EM burst from a power line will occur when the energy strings of the both energy fields have enough energy to actually break free from the electron system.

Specifically this means the following: the additional energy strings originally added to the electron system will shift from driving the motion to being simply the extending energy fields.

There must be enough energy added to the system to begin with, followed by enough energy diverted from the motions to the fields. There must also be enough magnetic energy strings converted from the magnetic strings to electrical strings such that both fields have enough launch energy.

This is when the power line will actually emit a burst of electromagnetic energy.

20. A specific burst of EM energy (specific inherent energy and pulsation frequency) will be emitted only when the Threshold Percentage has been reached.

> This means that the amount of energy in the fields, relative to the total energy of the electron system, allows a particular grouping of energy strings to break free and become an independent entity.
>
> The specific "Threshold Percentage" will launch an EM burst of a particular inherent energy and frequency of pulsation.

Power Loss from EM Bursts into the Air

Overview of Process for Power Loss

This also explains one form of power loss on a power line: where some power is lost to the air.

The purpose of a power line is to transmit power across the wire. This is done by pushing trillions of electrons along the wire. However, at the same time, most power lines will emit low amounts of EM bursts. These EM bursts are created because some of the energy for each electron has reached the threshold percentage, where enough energy has been diverted into the energy strings for launch of an EM burst.

When this happens, the electrons remaining have less energy. Why? Because some of the energy was just taken away, launched off from the electron. Consequently, we have two results: a) EM bursts are emitted, and b) the remaining electrons have less energy. Thus we have power loss along the wire.

Large Power Loss from Many EM Bursts

Most power lines will emit low amounts of EM bursts. However, sometimes a power line will emit larger quantities of EM bursts, and this is an indication of one type of power loss. Now we can understand why.

In this situation, much of the energy is going into creating and strengthening energy fields, rather than to delivering electrical power as desired. Thus, for any amount of energy at one segment of line, there is more energy to the energy fields, and less energy to the electrical power.

Soon the Energy Percentage Threshold is reached, and these fields take off as bursts of electromagnetic energy.

In turn, this means that some of the energy has gone away, specifically into the EM burst. Therefore, when we see many bursts of electromagnetic energy coming from a wire, we know that this wire is losing much of its energy to the air.

Power Line versus Antenna

On the other hand...we are creating EM bursts which we may be able to use to our advantage. This is the basic process of the radio transmission antenna, as will be discussed in the next section.

Radio Transmission Antennas

Introduction

The radio transmission antenna is very similar to the power line. The operation is almost identical. The only differences are the frequencies of alternating current, and the energy percentages directed to each area.

Basic Process of the Transmission Antenna

The transmission antenna is long piece of wire. This could be a thin metal wire or a complete metal tower. (The metal tower can be considered a very large wire).

The transmission antennas are usually installed vertically. This contrasts with most power lines that are installed horizontally. Whether placed vertical or horizontal does not matter in the operation. However, the direction of the antenna will be a major factor in the direction which the EM burst will travel.

Alternating current is sent through this wire. In an antenna, the current flows up and down, up and down. While the electrical current is flowing up and down, an electric field is created. This electric field flows left, and right, left and right.

At this point, everything is identical to the situation described above in power lines. We have alternating current, which then creates alternating energy fields. These alternating energy fields remain connected on the wire.

We keep adding energy to our power line. Some energy is diverted to the motions of electrons, while other energy is diverted to the energy fields. Eventually we reach the Energy Percentage Threshold, where the energy fields have enough energy to lift off. The energy strings break from their tether, and become independent entities. A burst of EM energy is created.

A transmission antenna and a power line are essentially the same devices. Both have alternating current traveling through wires, and both will emit EM bursts.

However, in an antenna EM burst are desired. In a power line, EM bursts are undesired, resulting in power loss.

Factors which Affect Frequency Emitted from Transmission Antennas

Introduction

Theoretically, any power line can emit any frequency of electromagnetic energy.

Frequency of Alternating Current and Frequency of EM Burst

The primary factor which determines the frequency of EM emitted is the frequency of the alternating current.

Remember that we are sending alternating current through the wire, and this alternating current produces an alternating energy field. This energy field naturally oscillates at the same frequency of the alternating current.

Therefore, the frequency of the alternating current we use is the primary factor which will determine the frequency of EM bursts created from a transmission antenna.

Height of Antenna

Another factor which determines the frequency of EM burst created is the length (or height) of the antenna. An EM burst emitted from an antenna will often have the same wavelength as the length of the antenna. For example, if the antenna is 1 meter tall, then we can produce an EM burst which has a wavelength of 1 meter.

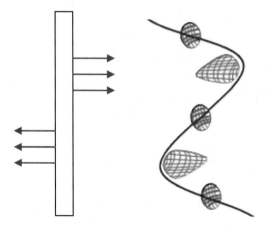

Antenna size = Length of Pulsation Cycle

Fraction Antenna Size

An advanced concept is this: using an antenna which is the fraction of the wavelength in order to create the desired burst of EM energy.

We can produce the same frequency (and same wavelength of pulse cycle) using antennas that are a fraction of the actual wavelength. Typical fraction lengths are ¼ and ½.

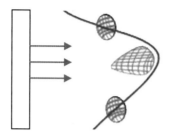

Antenna size = ½ Length of Pulsation Cycle

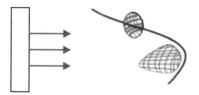

Antenna size = ¼ Length of Pulsation Cycle

How does this work? As long as we create the maximum amplitude, which will become the crest of our wave, at the right distance on our wire...then when the EM burst is launched, and then the EM Burst will take care of itself.

In other words, we do not have to go through a complete cycle, as long as we reach the maximum amplitude at the right length along the wire.

Remember how we create a burst of EM from a wire: We send alternating electrical current through the wire at the exact frequency we wish our emitted EM burst to be. The alternating current creates an alternating electrical field. The frequency and wavelength of this alternating electrical field will be exactly the frequency and wavelength of the EM burst. (However, these fields will stay attached to the wire until reaching the threshold percentage). When enough energy is diverted to the energy fields, then the EM burst will launch.

Note that we can also achieve this through wires which are fractions of the full size of wire, as long as our frequency of alternating current is the same desired frequency.

As long as the alternating energy fields reach a maximum at the desired length along the wire, then we can create the desired EM Burst. If the energy fields are launched at the moment where the energy fields are the maximum, then an EM Burst is created. As discussed earlier, this EM burst has a life of its own. It will expand and contract, it will pulsate, based on its own internal mechanism. Therefore, we do not need to go through an entire cycle on our wire in order to create the burst which has the proper frequency.

All that needs to happen in the beginning is that the maximum energy fields are allowed to exist along the wire at the appropriate distance on the wire (which will determine the wavelength and frequency of the burst).

Therefore, you can see that sending an alternating current through a wire, where the length of the wire is exactly ½ or ¼ of the energy field wave, will result in emitting the desired wavelength of EM being emitted.

Do note, however, that using the full length of an antenna, and allowing a full cycle of the energy field wave, will provide more efficient creations of EM bursts than the fraction antenna sizes will.

EM Bursts are Independent After Emitted

Overview

After the burst of Electromagnetic energy has been emitted, the burst is completely independent, and it has a life of its own.

More importantly, the mechanism of the EM Burst is completely different from the mechanism which emits the energy. Once in the air, the EM burst operates in a special way.

Mechanism in Brief

This mechanism will be discussed and explained in much greater detail in a later chapter. For now we will describe the process briefly: After the energy has been launched from the wire, the energy spreads out in a particular direction. Yet the flow of this energy is limited by a type of gravitational pull. This gravitational pull both determines the size of the EM burst, and is the cause of the contraction. The energy contracts inward, and then outward again, but in the opposite direction. Again the energy flows…but is limited by the gravitational pull. The energy then contracts, and the process repeats. (Full explanations, with drawings will be found in later chapters).

Creation and Perpetual Cycles of the EM Bursts

For now, the important concepts to really understand on the creation and perpetual cycles of the EM burst are:

1. Electrical and Magnetic energy fields start as being attached to an electron. Additional energy strings are added to the existing energy strings. Some additional energy strings go into the motions of the electron, while the other energy strings become attached to the field energy strings.

2. When enough energy has been diverted to the energy fields (the Threshold Percentage) then the energy fields will launch off the wire.

3. Once the energy is launched from the wire, the EM Burst is independent, with a life of its own.

4. The EM burst will pulsate based on its own internal mechanism. The EM burst will continue to pulsate based on this mechanism.

5. The frequency of the EM burst is determined primarily by the frequency of the inherent energy. This is the energy of the field (the non-driver energy strings) in the electron when on the wire.

6. Other factors which affect the frequency of the EM Burst include the frequency of the electrical current and the height or length of the wire.

7. The direction of forward travel of the EM burst is determined by the direction of the energy field at the time of launch.

8. As long as the fields have been launched, an EM burst is created. It does not matter where in the cycle of alternating current the electron is, if the fields have been launched off the wire, the launched fields will become a burst of electromagnetic energy.

A Few Points on High Voltage and EM Bursts

Introduction

At this point I would like to provide a few additional points on high voltage power lines and burst of electromagnetic energy.

Higher voltage electrons are more likely to emit EM bursts.

If the voltage is very high then an EM burst is more likely. This is because there is more energy applied to the electron, and therefore more energy strings which can be diverted from motion to the fields. With greater amount of energy to begin with, it is easier for fields to obtain enough energy to launch.

Conversely, if the voltage is low then an EM burst is unlikely. This is because there is not enough energy to allow the energy strings to break free.

High voltage power lines also have strong fields.

Remember that voltage is energy of the electron, and the energy of the electron is contained in the energy strings. Therefore, an electron with greater voltage naturally has stronger energy strings.

As discussed earlier, some of these strings create the motion, and some strings extend outward as energy fields. The strong driver strings will make the electron move faster, while the strong fields will be measured and felt more significantly.

A stronger energy field can be a thicker string or a longer string. If a thicker string, and you are close, then your measuring device (or your own electrons in your body) will register the energy field quite strongly.

Or, if the energy field is long, this means that there are very long energy strings. Some of these strings will extend hundreds of feet. You can stand far away and still detect them.

Also remember that the individual strings can vary in length. For example, a set of electrical energy strings can have lengths which range from 10 feet to 500 feet. You will feel all of the strings at 10 feet, but only a few of the strings at 500 feet. This is why the fields are stronger as you are close, and weaker as you step further back.

Amperage versus Voltage

There is a difference between Amperage and Voltage. Amperage measures the number of electrons involved in electrical current. Voltage measures the amount of energy of those electrons.

Amperage is the number of electrons moving forward at the same time. In other words, Amperage is a measure of electrical current.

We can increase the amount of amperage by using larger wires. With larger wires, we have more electrons per diameter of wire, and thus more electrons which can flow in a forward motion at the same time. The magnet is extending its magnetic strings all the time, and if more electrons are in the path (because of larger diameter wire) then those strings will be able to push those electrons and start the current.

Voltage is the overall energy of trillions of electrons on one small section of wire, at one moment in time. We can increase the voltage by using a stronger magnet. However, it is more common to use a transformer.

Transformers Increase the Voltage

The most common way to increase the voltage of the electrons is to use a transformer. I point this out because most high voltage power lines carry high voltage electrons based on the transformer, and have nothing to do with the magnets in the generator.

Increasing the Voltage for Different EM Bursts and Greater Emission

When we increase the voltage of the electrons we actually are able to do three things:
   1. Greater chance of EM bursts being emitted
   2. Stronger signal or greater intensity of EM bursts emitted
   3. Emit EM bursts of different frequencies

All of these effects are related to the same general concepts: we are adding more energy strings to the electron systems. This allows all of the above to occur.

1. <u>Greater chance of EM bursts being emitted</u>

As discussed above, when we add more energy strings to an electron some of those energy strings will add to the energy fields. When the energy strings in the field have enough energy to break free from the electron these strings will launch as an EM burst.

Thus, by adding energy strings we are both increasing the voltage, and increasing the chance that an EM burst will be launched.

Also note that this process for the burst emission from a single electron is essentially the same as done from a molecular vibration (discussed next chapter).

2. <u>Stronger signal or greater intensity of EM bursts emitted</u>

When we add energy to a wire we are not adding energy to just one electron. Rather, we are adding energy to thousands or millions of electrons at the same time.

For example, when we use a magnet in a generator, that magnet has a broad magnetic field. Those energy strings (of the magnetic field) will reach across the entire diameter of the wire. Therefore these energy strings reach all of the electrons in the wire.

At this point, whether it is a power line or a transmission antenna, the process is the same. When we "increase the voltage" we are in fact imparting more energy strings...not just to one electron, but to the millions of electrons across the diameter of the wire.

Of course each of these electrons transfers its energy to neighboring electrons across the lattice, essentially creating thousands of mini electrical currents down the wire.

Therefore we have millions of electrons up and down the wire, with varying amounts of energy strings. If we add a huge amount of energy strings at the beginning ("a much higher voltage") then there are many electrons which have enough energy to emit EM bursts.

Net result: increasing the voltage will increase the number of EM bursts emitted at the same time. In a radio antenna, this is known as a stronger signal. For energy beams or light, this is known as greater intensity.

3. Emit EM bursts of different frequencies

As discussed elsewhere in this book, in order to emit different frequencies of EM bursts we must have groups of energy fields that differ in total energy (inherent energy), combined with the launch energy to cause that group of strings to lift off.

Consider this simply with amount of material. In order to build different size rockets, we need different amounts of material. And, the largest rocket we can build will depend on the maximum amount of material we have. Similarly, the highest energy EM burst we can emit will depend on the maximum amount of energy we add to the electron system.

Thus, if we increase the voltage, we are in a sense giving the electron the opportunity to build a higher energy EM burst.

Furthermore, increasing the voltage to large amounts will give us more options as to which EM bursts will be emitted.

Again, consider the material for a rocket. If we are given an enormous amount of material, we can build multiple rockets. We can build several small rockets. We can build a few small rockets and one larger rocket. Or we can put it all into one very big rocket.

In the same way with energy, when we add an enormous amount of energy strings to the electron, that electron has more options in the grouping of energy strings to be launched.

Finally, remember that there are millions of electrons with varying energies at the same time. Therefore, it is possible for each electron to emit a different frequency EM burst at the same time.

Note that the options are usually limited, based on frequency of electrical current and length of the wire. However, there can be a few options of EM frequencies being emitted from any one electron, and more possible options if we transfer greater energy to each electron. Thus, when we apply large amounts of voltage to our wire, it is often possible for a few different frequencies of EM bursts to be emitted along the wire at the same time.

Summary Points regarding Voltage

Regardless of how the voltage is created, whenever we increase the voltage of the electrons in our wire there is a far greater chance of EM bursts being emitted from that wire.

In addition, when we increase the voltage of a wire we are adding energy to many electrons at the same time. Therefore, there will generally be more electrons emitting bursts of electromagnetic energy at the same time.

If the frequency of the electrical current and the height of the wire are specified and constant, then these bursts will generally be of the same frequency pulsation. This creates a stronger radio transmission signal or a greater intensity of light. Otherwise, the frequencies of EM bursts may be random.

Also, with greater voltage we have different options for EM burst being emitted. First, the highest energy (and highest frequency) of EM burst emitted can only be as high as the additional energy added to the system (and usually a bit less).

Second, if we add large amounts of energy this gives the electron more options for assembling different groups of energy strings, thereby emitting several smaller EM bursts one after the other.

In total, the amount of energy we add to an individual electron or to a group of electron will have a significant effect on the types and amounts of EM bursts being emitted from any location.

Chapter 9
Creation of EM Bursts: Molecular Vibrations

Introduction
In this chapter we will look at the creation of electromagnetic fields from vibrating molecules. Each of these sources use the same basic mechanism for creating bursts of electromagnetic energy. As stated in the previous chapter, most EM bursts are created in the same way. The mechanism requires:
1. Alternating electrical current, and
2. A high enough energy to meet the threshold for emission.

Molecular Vibrations Creating EM Bursts: General Overview

Introduction
The other most common way to produce bursts of electromagnetic energy is through the emissions of molecular vibrations. We will see that the creation of EM bursts from molecular vibrations is very similar to the mechanism previously discussed for power lines and transmission antennas.

Basic Process of Molecular Vibrations and EM Emission
All molecules vibrate. Most of the time they vibrate at the lowest energy level. However, when energy is absorbed by the molecule, much of that energy can be diverted to the molecular bonds. Now these molecular bonds vibrate with much higher energy.

These molecules will vibrate with this higher energy for a while, but these molecules prefer to be at a lower state. Therefore the molecules get rid of this extra energy. One of the methods to get rid of this excess energy is by emitting a burst of electromagnetic energy.

Advanced Understanding of Process
In the following sections we will come to understand this process more clearly. We will see that the molecular bond is similar to the power line, and the creation of EM bursts here is similar to the processes described earlier.

Current and Fields in Molecular Bond Vibrations

Introduction

We will start with a basic molecular bond. A molecular bond is a state of existence where two atoms are physically connected, typically through the sharing of electrons.

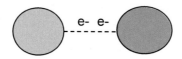

The Molecular Bond as Power Line

A molecular bond is typically drawn with a straight line. There isn't a physical line in reality, but it helps to show molecular bonds. I mention this point because I would now like to take our imaginary bond line, and pretend that it is a power line. We do this to see the similarities to the mechanisms described above.

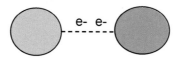

Molecular Vibrations

What is a molecular vibration? A molecular vibration is where the atoms in the molecular bond expand and contract, expand and contract. The atoms do this repeatedly. That is a molecular vibration.

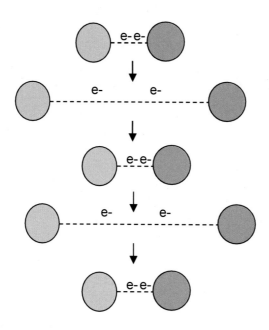

Electrical Current in Molecular Bond Vibrations

The electrons in a molecular bond actually create a type of electrical current. For example, in the "expansion" stage:

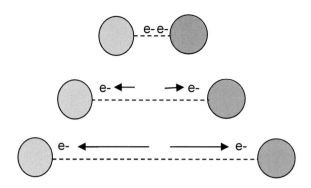

You will notice that as the atoms spread apart, and as the electrons move further apart, the electrons are in fact moving. The electron on the right is moving further and further to the right. The electron on the left is moving further and further to the left.

Now remember this: any time an electron moves in a line, current is created. Electrical current is electrons in forward motion. Indeed, that is what we see here as the molecular bond expands.

Alternating Current in Molecular Bond Vibrations

Furthermore, we see the same thing when the molecular bond contracts. Again the electrons are moving in a linear direction, which by definition is the existence of current. The only difference is that each electron is moving in reverse direction (toward contraction) as compared to its first direction (toward expansion).

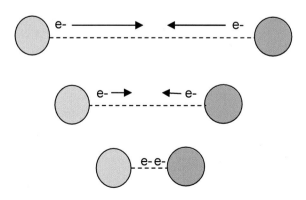

Therefore, the molecular bond, by expanding and contracting during vibration, also creates a type of alternating current.

Energy Fields in Molecular Bond

Because the vibrating molecular bond creates a type of current, the vibrating molecular bond also creates alternating energy fields.

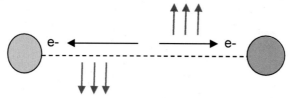

In addition, the electrons have several other motions (including electron spin and electron vibration). These motions produce energy fields (both electric and magnetic fields).

Taken together, the variety of energies and motions of an electron in a molecular bond will produce alternating electric fields and magnetic fields as the molecule vibrates.

These alternating energy fields are what will create the EM burst, once the Energy Percentage Threshold has been reached.

A More Accurate View of Molecular Bonds: Molecular Bonds as Parallel Wires

Introduction

At this point we will illustrate a more accurate picture of molecular bonds. These models will more accurately show the nature of molecular bonds, the expansion and contraction, and the creation of alternating energy fields.

The first new model will be that of parallel wires. The second new model will be an oval racetrack.

Molecular Bonds as Parallel Wires

A more accurate depiction of molecular bonds would be two parallel wires. Our first model had only one line, and both electrons are located on that line. In this model we have two parallel lines, and each electron is on a separate line.

Therefore, imagine that these two electrons exist on two separate yet parallel wires.

Position A for electrons in parallel wires

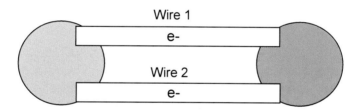

Electron Movement in Parallel Wires / Bonds

The electrons are not only on different wires (different bonds), but they move in opposite directions. Watch the process as each electron moves. Watch what happens as the current flows.

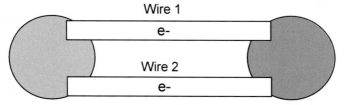

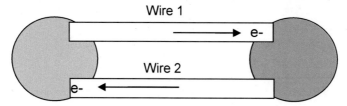

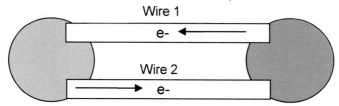

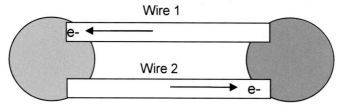

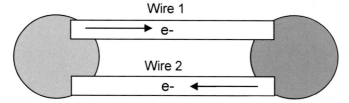

78

Expansion and Contraction

The first thing to notice is the expansion and contraction of the molecular bond. Each electron is moving in the opposite direction with respect to the other. This is the situation at all times. Therefore, the net result is an expansion, followed by a contraction, over and over again.

For example, position B shows both electrons moving outward:

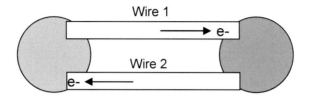

This results in electron 1 pushing to the right, and electron 2 pushing to the left...and therefore results in an overall expansion of the molecular bond. Position B would be more accurately drawn as:

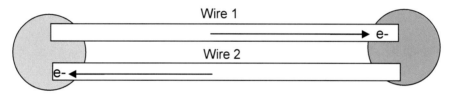

Similarly, Point C shows both electrons moving inward. This results in an overall contraction of the molecular bond. (The molecular bond contracts back to its original length).

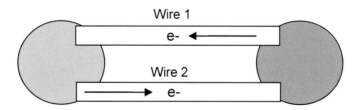

Current Flow and Energy Fields of Each Wire / Molecular Bond

The next thing to notice is the current flow and energy field of each wire (of each one of the parallel molecular bonds).

Remember that these wires are independent. Therefore it is best to look at each wire individually.

We will begin by looking at wire 1 only, as it goes through its cycle of alternating current.

When we first see the electron move, the electron moves to the right. Therefore the current flows to the right. At the same time, this creates an energy field upward:

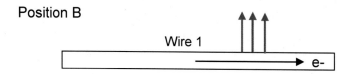

When the electron reaches its furthest point, it turns around and heads the other direction (Position C and D). Therefore, the current flows to the left. At the same this creates an energy field flowing downward.

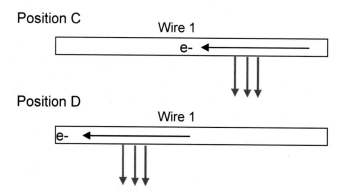

The electron then turns around, and travels the other way. This appears very similar to the first diagram above.

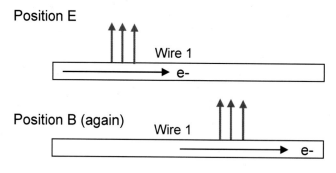

You can see that the electron in wire 1 (the electron in molecular bond #1) naturally creates an alternating current, and consequently it creates an alternating energy field. This electron, on its own path, is therefore capable of creating a burst of electromagnetic energy.

Similarly, electron #2 on molecular bond #2 performs the same actions, just in the opposite direction with respect to electron and molecular bond #1. (See below for diagrams).

Therefore, each wire (each molecular bond) naturally creates an alternating current, and an alternating energy field. Consequently, either electron, in either molecular bond, is fully capable of producing a burst of EM energy (once the threshold has been reached of course).

Side by Side Comparison: Activities in Each Wire at the Same Time

The following sets of drawings show the electrical current and energy fields which exist in each wire (molecular bond) at the same time.

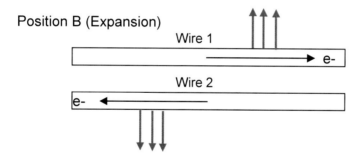

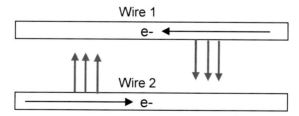

Position D (Expansion, electrons pushing opposite sides from previous expansion).

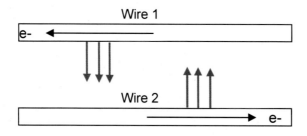

Position E (contraction)

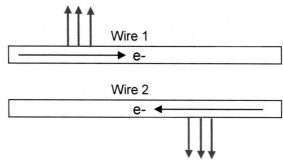

Model #2 of molecular bond:
Each molecular bond is a parallel wire,
where each bond operates independently.

Electrons in each bond travel back and forth,
thus creating alternating current and alternating energy fields.

Molecular Bonds as a Racetrack:
The Elliptical Looping System
aka "The Molecular Orbital"

Introduction

A more accurate depiction of the molecular bond, and perhaps the most accurate model today, is an elliptical looping system – much like a racetrack.

Imagine a racetrack in the shape of an ellipse. On this race track are two cars. Each car is positioned equal distance from the other. We see the cars travel round and round the track, in a never ending loop.

In a similar way, the molecular bond between two atoms is an ellipse. This molecular bond is like the racetrack, and the electrons are like the cars. Just as the cars go round and round the track, always equal distance from each other, so the electrons go round and round the orbit, always equal distance from each other.

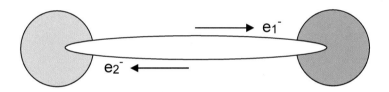

Seemingly Opposite Directions

Because the cars are equal distance apart, we notice that their motions seem to be opposite. As the first car travels right on the straight section, the second car travels left on its straight section. As the first car turns south, the second car turns north. In this way, the actions of the cars always appear to be opposite.

The electrons in their molecular orbit perform the same traveling pattern as the cars on the race track. The electrons are both on the same loop, yet their relative distance apart gives them the appearance of always acting in opposite directions.

The Molecular Orbital

I call this system "The Molecular Orbital". It is more than a simple bond. In this model the electrons travel in an orbital path. Each electron travels through one atom, around the nucleus, to the other atom, around its nucleus, and back again. It is a continuous path, a continuous loop, which joins both atoms together.

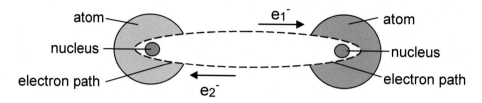

The Molecular Orbital System

Motion of the Electrons in the Molecular Orbit

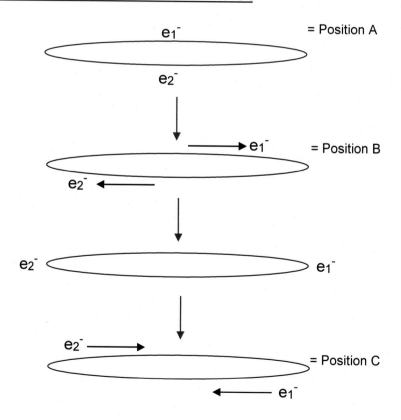

Note that this latest drawing is equivalent to Position C because electron 1 and electron 2 are moving in the opposite direction as they were in Position B. However, our more accurate model shows that each electron is on a loop, rather than going back and forth on the same wire.

Therefore, each electron is traveling on the other linear portion of the track (rather than back on the same wire). This continues with Position D below.

The effect is the same as in the previous model, but it is a more accurate description of the process.

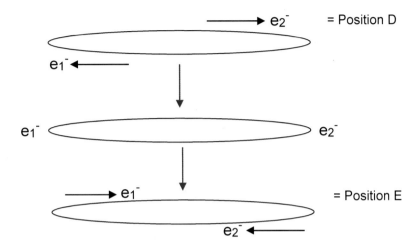

Expansion and Contraction with Looped Molecular Orbit

We can see that these electrons in the looped orbit perform the same operation as if the electrons were on separate wires (as in the earlier model). First we can see that the molecule expands and contracts due to the relative motion of these elections:

☐ 1

Molecular Orbit, Position B:

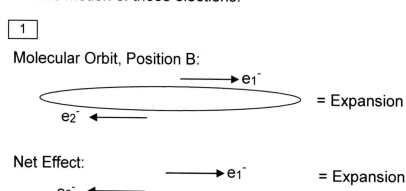

☐ 2

Molecular Orbit, Position C:

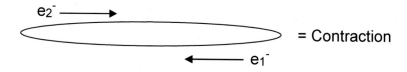

Model #3 of molecular bond:

The molecular bond is actually a molecular orbital, much like a race track. It is an ellipse, it is a closed loop.

The electrons travel are equal distance apart, which makes them appear to travel in opposite directions. Their combined motions around the loop creates the expansion and contraction of the molecular bond.

Creation of Energy Fields in Looped Orbit

The looped orbit also creates the desired energy fields. Notice that for each electron, the electron travels one direction, and then the other. Although the electron is not on the same line (as in the earlier model), the electron *is* going the opposite direction. Therefore, the electron creates alternating current.

Also notice that because each electron creates alternating current by traveling in this ellipse, each electron will also create alternating energy fields.

Therefore each electron, by going through this elliptical orbit, will create the alternating energy fields which can become emitted electromagnetic energy bursts.

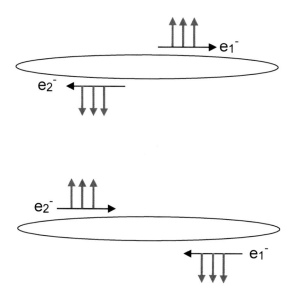

The key is to place the electrons in the right place relative to each other before you begin. Note that this naturally occurs, because the electrons like to keep away from each other. In the same way that birds on a wire naturally space themselves apart, so to the electrons will naturally space themselves out, regardless of how many electrons are in the orbital.

If the electrons are spaced apart appropriately, then you can create the same effect of the parallel wires (when the electrons are on the straight part of the oval). You will also create the same expanding and contracting, and yet have this racetrack operation....like cars continuously traveling around the race track.

Each electron travels along two straight portions of the loop.

When the electron travels the linear portion from left to right, an electric field is created flowing upward. When the electron travels the other linear portion, now traveling from right to left, the electrical field is created flowing downward.

Therefore, as the electron travels the orbit alternating electrical fields are created. These alternating electrical fields will become the basis for EM bursts (once the threshold percentage is reached)

Two Molecular Orbital System

Overview

In actuality, two electrons never travel on the same loop. Each electron will always be on a separate path. Therefore, for what we commonly know of as the 2 electron molecular bond is actually a two molecular orbital system.

The Two Molecular Orbits

Thus we have two molecular orbits. One electron is on each orbit. As illustrated above, each of these molecular orbits is a path where the electron travels through the interior of each atom, around the nucleus, and to the interior of the other atom.

Yet what is the same is that each electron travels in the opposite direction. For example, electron 1 will travel on its molecular orbital from left to right, while electron 2 will travel on its molecular orbital from right to left.

The Two Molecular-Orbital System

You can see in the illustrations how the two molecular orbital system shows the following:
1. Molecular Bond
2. Creation of Molecular Vibration
3. Cyclic Energy Fields

1. Molecular Bond

Each electron travels in a loop through both atoms. This is what actually connects both atoms together.

A second electron on a second molecular orbital does the same thing. This second electron travels through the same two atoms (just on a different orbital path, and traveling in a different direction).

Thus two electrons traveling through both atoms will connect those atoms to a much greater degree. In other words, this makes a stronger molecular bond.

2. <u>Creation of Molecular Vibration</u>

The two molecular orbital system also creates the molecular vibration. Each electron may operate independently, yet together the net result is the expansion and contraction of molecules.

Specifically, as both electrons move closer to their opposite atoms, the atoms stretch further apart. Then as both electrons travel in the region between the atoms, the atoms move closer together. This produces the repeated expansion and contraction of the atoms, more commonly known as the molecular vibration.

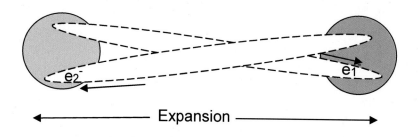

←――――――― Expansion ―――――――→

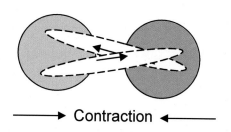

―――→ Contraction ←―――

3. Electrons in a Cyclic Loop with Cyclic Fields

Each electron travels in an orbital, which is a continuous loop. Therefore, each electron exists in a cyclic path. This cyclic path produces cyclic electrical fields, cyclic wave patterns, and eventually cyclic waves of electromagnetic energy.

Cyclic Alternating Current

Let us look at one electron, such as electron #1. We begin with electron #1 traveling to the right. As the electron travels, it creates current. (By the very definition of current, any electron in forward motion creates a current).

Then as the electron travels around the nucleus toward the other atom, the electron is in fact traveling the opposite direction. Therefore, the electron is creating an alternating electrical current as it travels.

Cyclic Alternating Energy Fields

The energy fields, particularly the electric fields, are also cyclic and alternating. In general, when the electron is moving from left to right, the electrical energy strings flow upward. When the electron is moving from right to left, the electrical energy strings flow downward.

We can see this on our molecular orbital, as the electron travels left to right on one straight segment of the path the electrical energy field extends upward.

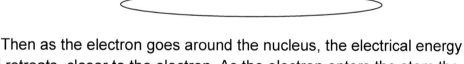

Then as the electron goes around the nucleus, the electrical energy field retreats, closer to the electron. As the electron enters the atom the energy field shrinks, until it is almost non-existent.

And as the electron travels right to left on the other straight segment of the path, the electrical field extends downward.

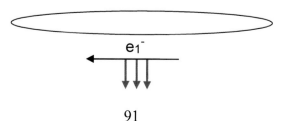

Therefore, when we have an electron traveling in an elliptical orbit, we will have alternating energy fields. These alternating energy fields will continue repeatedly as the electron continues to travel the loop of the molecular orbital.

Process of Molecular Vibrations Creating EM Burst (looking at the Path of One Electron)

Introduction

We are now ready to look at the specific process of creating a burst of electromagnetic energy from a vibrating molecular bond.

The Molecular Bond as Elliptical Molecular Orbit

For our model of the molecular bond we will use the most accurate new model: The Set of Molecular Orbitals.

Let us review the main components of this model before we begin discussing the creation of EM bursts.

A. The basic configuration of this model is two continuous loops. Each loop passes through both atoms. There is one electron on each loop, each electron travels through both atoms, and each electron travels in opposite directions.

B. The "bond" is created because the electron travels through both atoms. Thus, both atoms share the same electron, creating that bond. And if there are two molecular orbitals, then we essentially have two distinct bonds, which essentially doubles the strength of the connection between the two atoms.

C. The "vibration" is created because the electrons travel in opposite directions on their respective orbitals. As each electron enters the regions of their respective atoms, the atoms are pushed further apart, thus creating expansion. As the electrons start to head back, the atoms are closer, and this creates contraction.

The rate of frequency for the vibration is related to the speed of the electrons, and the speed of the electrons is related to the energy of the electrons. Therefore, as the electrons acquire greater energy, the molecular bond will vibrate faster.

D. The "fields" of the electrons are strings which extend into the air from the electron. In normal energies of the electrons, these energy strings remain attached. However, given enough additional energy (absorbing energy strings from an external source) some of these field strings will break free, thereby creating a burst of EM energy.

Only One Electron, One Orbital, Needed to Produce EM Burst

We only need one electron to emit a burst electromagnetic energy. Therefore we only need to look at one molecular orbital when discussing the process of EM burst from a molecular vibration.

Begin with Energy Fields Attached to the Electron

In the stable state of the electron in its molecular orbital, the electron has a certain amount of energy. This energy is observed as a particular amount of motion, and a particular strength of the energy fields.

The energy strings which create the motion of the electron and the energy strings which create the energy fields can and do change roles throughout the life of the electron in that orbit.

If enough energy is diverted to the energy strings of the energy field, then those energy strings can break from the electron, and become an independent EM burst.

However, this usually does not happen. Usually there is not enough energy in the electron's stable state for this to occur. Therefore the energy strings remain attached to the electron.

Adding Energy from an External Source

The next step in the process is to add energy from an external source. For molecular bonds, this is most commonly done by the electrons absorbing electromagnetic energy.

There are numerous photons of electromagnetic energy flying around us at all times. Some of those photons will hit a molecule and be absorbed. The exact process of absorption will be discussed in greater detail in a later chapter. Briefly: the energy strings of the photon will join the energy strings of the electron. Thus, the photon "disappears" while the electron has obtained all of that energy.

Now the electron in the molecular bond has additional energy. The amount of additional energy is the exact amount of energy in the photon absorbed.

Increasing the Frequency of Vibration

Now that the electron in the molecular bond has all of this additional energy, the electron will have greater speed and stronger energy fields.

Regarding the speed: some of the energy strings absorbed will be driver strings. Thus, with more driver strings, the electron will move much faster. This includes faster forward motion. And because of the path of the molecular orbital, this additional energy as driver strings will cause a faster vibrational frequency.

Increasing the Strength of the Electrical Field

Some of the absorbed additional energy strings will become field strings. Thus the field will become stronger. In addition, as we have discussed previously, driver strings will become non-driver strings throughout the life of the electron.

Therefore, for both these reasons, the non-driver energy strings (the fields) of the electron after absorbing energy will be much stronger.

If at any time those energy strings have enough energy to break free from the electron, then a burst of electromagnetic energy will be emitted.

Launching the EM burst from the Electron

A burst of electromagnetic energy will be emitted from a molecular bond when the energy field strings have enough energy to break free from the electron.

Most of the time, these alternating energy fields will stay connected to the orbit (like the dog tethered on a leash). However, when enough energy is imparted to the energy fields, the energy fields will lift-off as an independent burst of EM energy.

Also remember that the total energy of the molecular orbit system is diverted into two areas: the motions of the electron, and the energy fields. In order for alternating energy fields to take flight as an EM burst, there must be enough energy diverted to the EM fields as compared to the total amount of energy in the orbit system. (This is the threshold percentage). At this point, a particular frequency EM burst is created.

This is the mechanism in which a burst of electromagnetic energy is created from a vibrating molecular bond.

Possible EM Bursts to be Emitted

It is possible for one of several frequencies of EM bursts to be emitted. If the molecular bond can emit several frequencies of EM bursts, the particular frequency emitted will depend on:
1. The amount of energy absorbed
2. The first Threshold Percentage reached

The details will be discussed later. In brief: remember that earlier we discussed inherent energy and launch energy, and that the Threshold Percentage is combination of these. Thus, when the energy strings are arranged in a particular way, and they have proper amount of energy for lift-off, then that is the group of energy strings (and hence the energy of the EM burst) that will be emitted.

Reduction in Frequency of Vibration after Emission

As a consequence of the EM burst emission, the molecular bond vibrates at a slower frequency.

This is easily understood: the electron has fewer energy strings overall because some energy strings have left as the EM burst. Thus, there are fewer driver energy strings to drive the electron forward. Consequently, the electron will travel slower, and the expansion and contraction process will be slower.

Therefore, after the molecular bond has emitted an EM burst, the molecular bond will naturally vibrate slower.

Note: "Subsequent Emission"

Notice that in this process there are two photons. The first one is the external photon. This comes from the air, hits the molecule, and becomes absorbed. The second photon is the EM burst which is emitted from the molecule after absorbing the additional energy.

In order to clarify the two during discussion, I will refer to the second photon, the one coming from the molecular vibration, as "subsequent emission."

Also, note that the energy of the subsequent emission is usually smaller than the energy of the incoming photon. (The subsequent emission cannot be greater than the incoming photon, and is rarely equal to it.).

Additional Factors to Note
in the Creation of EM Bursts from Vibrating Molecule

Now that we have discussed the basic mechanism for creating an EM bursts from a vibrating molecule, it is time to discuss a few related concepts. The first set of these concepts are:

1. Frequency of EM burst is related to speed of electron.
2. Loss of energy in an orbital system is diverted to energy strings and EM burst.
3. Amount of time a molecule exists in a higher energy state before emitting EM burst is variable, depending on the threshold percentage.
4. When an orbital system acquires energy, this system can exhibit higher energy through change in speed, shape, and size.
5. Most molecular bonds can emit one of several frequencies of EM burst; which one emitted depends on the relative percentage of energy in energy fields versus energy in the total orbital system.

Each of these concepts will be described in separate sections below.

Frequency of EM Burst is Related to Speed of Electron

The speed of the electron in the orbit can be correlated with the frequency of the EM Burst. Sometimes there is a direct cause and effect, but more often it is more of an indication for other processes. There are several reasons for this:

1. More Driver Strings may Become More Field Strings

Remember that the electron on the elliptical orbit is like a car on a race track. On a race track we can make the car go faster and faster and faster. Similarly, we can make the electron go around the loop faster and faster and faster.

Then remember that the speed of the electron is caused by driver energy strings. Therefore, when we have more driver energy strings, the electron will move faster.

At the same time, remember that driver strings migrate to merge with the field strings. If there is enough energy to break free, then the EM burst will be emitted.

Therefore, we have a correlation: adding more energy strings as driver strings will first make the electron move faster, and then cause the EM burst to be emitted.

2. <u>Correlation between faster speed and overall energy of electron</u>:

Whenever the electron is moving faster, this means we have more driver strings. Yet, this is actually an indicator of the total energy. In other words, if the electrons have more driver strings, it is actually likely that the electron will have stronger energy field strings, as well as more free flowing strings in the middle.

Consequently, stronger fields and more energy strings will result in higher energy (and higher frequency) EM bursts being emitted.

Therefore, when the electron travels faster, this is actually an indication of more energy strings in the system, which can produce higher frequency EM bursts.

> The rate at which the electron travels around the molecular orbit is an indicator of the strength of the internal energy strings, and therefore correlates with the frequency of the EM burst which may be emitted.

Loss of Energy in Orbital System When Burst Emitted is Due to Percentage of Diverted Energy

When energy is emitted as an EM burst, the overall energy of the orbital system will become much lower. Now we can understand all the interrelated factors which explain each part.

Remember that the total energy in the Orbital System is a combination of energy in the electron motion, and energy in the energy fields. That is a total amount for the whole orbital system. The percentages to each can be any combination, but their combined energies will reach the total.

When enough energy is diverted to the energy field to take flight as an EM burst (let's say 60% of the total) then what remains of the original total is only 40%. Consequently, the orbital system has only 40% of the original amount of energy, which is a much lower amount.

These interrelated concepts explain: a) how the EM burst obtains its amount of energy from the orbital system, and b) how a molecular bond loses energy when an EM burst is emitted, c) resulting in the orbital system existing at a lower energy.

Amount of Time in Higher Energy Before Emit

The amount of time which the orbital system (electrons and energy fields) stays in a higher energy state depends on the threshold percentage.

The electrons and energy fields of the orbital system can continue to have higher energy for a long time, or for a short time. There is no fixed time for any orbital system to be in its higher energy state. Also, the amount of time is completely independent of the energy of the system.

The only factor that matters is the Threshold Percentage. Only when the Energy Percentage in the energy field reaches the threshold amount will an EM Burst be created.

This could take seconds or minutes or an instant. All that matters is that enough energy is diverted to the energy field to create lift-off. Only when the random combination of other events diverts enough energy to the field will the EM burst be created, and the electron revert to lower energy. Until that time, the orbital system will remain in that higher energy state.

Higher Energy Orbitals:
How an Orbital System Exhibits Acquired Energy

Introduction

When an electron (and the corresponding energy field) acquire more energy, that energy will change the nature of the orbital system. There are several variations of how this new energy state may be observed.

Speed of Electrons

The first way in which the orbital system may change is the electrons may increase speed. Remember our analogy of the car on a racetrack. This car can increase its speed, traveling faster and faster around the loop. The same happens with the electron on the orbital. When energy is added to the orbital system, the electron can travel faster around the orbital loop.

Therefore in many atoms and molecules, the electrons in the orbital systems will move faster as a result of acquiring additional energy.

Shape of Orbital

Another way in which the orbital system may change with additional energy is through the shape of the orbital.

The orbital may be stretched or may be shortened. The angle of the orbital may change, such as 15 degrees diagonal from the original orbital.

In addition, the electron may choose to create a bizarre shape, performing many turns while completing its loop. (Think of a complex race track with many turns. The car travels a closed circuit, but takes many turns before getting to the starting point again).

In general, many shapes are possible. However, for each atom and each molecular bond, there will be only a few options of orbital shapes.

Therefore, the orbitals of many atoms and many molecular bonds have multiple options for orbital shapes. The specific orbital shape is how the orbital system will often change when acquiring additional energy.

Schematic of EM Absorption and Subsequent Emission

Introduction

Before we get into further details regarding why an electron will emit one frequency of EM burst rather than another, it is important to understand the traditional schematics of energy changes for an electron.

We will review the basic schematics used for energy absorption and energy emission. Then we will discuss how these correspond to real properties of the electron, particularly the arrangements of energy strings.

Traditional Schematic for EM Absorption and Emission

In the traditional schematic of EM absorption and subsequent emission, the energy levels are drawn as a series of lines, similar to a series of book shelves.

Level 4 ———

Level 3 ———

Level 2 ———

Level 1 ———

Each level of the schematic usually refers to the possible energies which can either be absorbed by the electron or emitted by the electron.

a. The higher "shelf" represents a higher energy level.

b. The "shelf" or "level" system exists because the electron will tend to absorb only certain frequencies, and emit only certain frequencies.

c. The spacing between each level is an approximately equal to the amount of energy between each level.

Clarifying the Meaning of the Schematic in Each Case

The traditional schematic of EM absorption and subsequent emission can be used to represent a number of different situations. Note that the author should clarify exactly what the levels are referring to in his diagram:

a. The author should specific which electron the schematic refers to, because every electron in an atom (or every molecular bond in a molecule) will have a different set of energy levels in the schematic.

b. The author should also specific whether the schematic refers to the absorption levels, the emission levels, or both.

Single Schematic versus Two Schematics

A single schematic is commonly used to show how electrons can absorb certain energies, then emit certain energies.

However, in reality there are two sets of schematics: 1) the absorption schematic, and 2) the emission schematic. Two schematics really exist because electrons tend to absorb only certain energies, and tend to emit certain energies. These sets of energies are not necessarily the same.

Yet for simplicity in understanding the processes and making comparisons, it is useful to merge these two schematics together. This means all levels are shown, for absorption and for emission, in one schematic, though the "emission" levels are not likely to be "absorption levels", and vice versa. They are simply drawn in the same picture for story-telling and for simple comparisons.

Levels in the Schematics

The lowest level usually represents the energy level where the electron is stable. The electron does not emit any EM burst on its own. Note that the energy level of the lowest level will be different for each electron in an atom.

The higher levels represent higher energy levels. In an absorption schematic the higher levels represent the levels to which an electron will absorb energy.

The spacing between each level is approximately equal to the difference in actual energy levels.

Electron in This Orbital versus Electron in the Next Orbital

Notice that the highest level in any schematic is usually the highest level of energy for that electron in that orbital path *for which the electron will be unstable*. Any additional energy will send the electron to the next orbital. If the electron gains enough energy, it will reach that next level, where it will reside in the next stable orbital.

Thus, beyond the highest level of any one schematic is another schematic, for a completely different orbital.

For example, adding enough energy to send the electron's energy to level 4 will be the highest unstable energy state for that orbit. Yet, if we added enough energy to go beyond that, we have in fact moved the electron to a completely new orbital, and the electron is actually stable once again. (Further details will be discussed later).

Energy Level Schematic Understood with Energy Strings

Introduction

With our understanding of energy strings and the process of launching an EM burst, we can now correlate the schematic of energy levels with the actual processes related to energy strings.

Specifically, every energy level in the schematic, whether absorption or emission, corresponds to the total amount of internal energy within the electron system. This total internal energy of the electron consists of:

a. The total number of electrical energy strings
b. The total number of energy magnetic energy strings
c. The thickness and lengths of each of those strings

Taken together, the number of energy strings (and their dimensions) in the electron system will give us the total energy of the electron system

Absorption Schematic and Energy Strings

In the absorption schematic, each line represents the new energy of the electron after the incoming photon has been absorbed. We can understand this in terms of energy strings as follows:

When the electron absorbs a photon what actually occurs is that the energy strings of the photon enter the electron system. There the energy strings physically merge. Once this occurs, we have additional energy strings in the electron system.

Some photon energy strings are merged with the energy fields of the electron, while other photon energy strings are simply free flowing within the interior of the electron sphere.

Thus, the total amount of energy strings within the electron system has increased. Therefore, the internal energy of the electron has increased. This amount of energy is represented by one of the lines in the schematic.

Furthermore, each energy level in the schematic for absorption represents a different amount of energy absorbed. This corresponds to different energies (frequencies) of photons being absorbed.

Thus, when the photon has higher energy, it has more energy strings, and therefore when absorbed by an electron this will add more energy strings to the system. This will result in greater internal energy for the electron, which is then represented by a higher level line in the schematic.

Emission Schematic and Energy Strings

The schematic for subsequent emission of an EM burst can similarly be related to the processes of energy strings.

When an electron emits an EM burst, the physical process involves energy strings being launched from the electron into the air as independent entity. Therefore, when an EM burst is launched the electron has fewer energy strings than before. Of course fewer energy strings also means that the internal energy of the electron is much lower.

Thus, when an EM burst is launched, the remaining energy strings are much fewer, resulting in less internal energy of the electron. This is represented by a lower level line in the schematic.

Various Options in Emission Schematic using Energy Strings

The various options of energy levels in the emission schematic can now be understood in terms of energy strings and Threshold Percentage.

Every level in the emission schematic represents a different level of energy, and as we have learned the reality of this level of energy is the amount of energy strings within the electron system.

As stated above, when an electron emits an EM burst what is actually occurring is a group of energy strings merge together and launch from the electron. This results in fewer remaining energy strings in the electron, and therefore the total energy of the electron is lower. This process is represented on the schematic as an electron starting at a higher level (such as level 4) then drop to a lower level (such as level 3).

Yet there are various energy levels on the schematic, which represent various levels of energy to which the electron can "drop down". These can be understood by using the Threshold Percentage.

Whenever a particular group of energy strings merge together in such a way for those string to reach the Threshold Percentage, then those strings will launch as an EM burst. Yet there are several combinations of energy strings which may come together and provide enough energy to launch. Therefore, whichever of those combinations of energy strings is merged together first, then that is what will be launched from the electron.

Thus, every "arrow down" on the emission schematic represents a specific group of energy strings coming together, merging, and being launched.

And consequently, every line on the emission schematic represents a new internal energy of the electron, after emitting a particular EM burst.

How an Electron "Knows" What Frequency to Emit

Scientists have often wondered how an electron "knows" how to emit a particular frequency of EM burst, in order to drop down to a particular level of energy. We can now explain this fully.

First, the order of the processes as phrased in the question must be inverted. The electron emits a frequency of EM first, and then the new energy of the electron is a result of the energy lost to the air.

Second, the specific frequency of EM burst emitted will depend on what Threshold Percentage is reached first. Remember that all these energy strings are in the system, floating around. Also, as will be discussed in a later chapter, there are various motions which cause the energy strings to move in multiple directions. Therefore, these energy strings will combine, merge, rearrange, separate, and migrate. In brief, there are numerous puzzle pieces floating around in the container.

When these energy strings combine in a particular way, they will have enough energy to break free from the electron, and the EM burst will be launched. Yet, there are several possible combinations for EM bursts to be launched, and the motions of the energy strings are quite dynamic. Thus, whichever of the possible combinations of energy strings is created first, that is the particular EM burst which will be launched.

The remaining internal energy of the electron is thus the new energy state of the electron.

Therefore, the electron does not need to "know" how to emit a specific frequency in order to arrive at a particular energy state. Rather, the electron emits a particular EM burst depending on which the possible combinations of energy strings is created first. The remaining amount of energy strings in the electron then "drop" the electron to its lower energy state.

Which Energy or Frequency will be Emitted:

Which energy (and hence frequency) of EM burst will be emitted
(of the possible options for a given molecule)
will depend on how the energy strings in the electron
come together, arrange themselves, and merge.

The combination of energy strings which comes together first
(of the possible options)
will be the frequency of EM burst that is emitted.

Energy Percentage Determines
Which EM Burst (of possible options) will be Created

Introduction

As discussed above, many molecular bonds and many atoms are able to emit several frequencies of EM Bursts. Which frequency will be emitted (of the possible frequencies for that molecular bond) will primarily depend on how the energy strings will arrange themselves.

Yet we can also look at which frequency of EM burst will be emitted in the terms of Energy Percentage. Remember that the ability for an EM burst to lift off is not just related to the amount of energy in the field strings, but must also be a certain percentage of energy in the field strings *as compared to total energy of the electron system*.

As stated earlier, the exact Energy Percentage required for a group of EM strings to launch from the electron is the Threshold Percentage. Therefore, in the next several sections we will discuss three things:

1. The Threshold Percentage for different possible EM bursts in one electron system.

2. The Threshold Percentages as calculated from different starting points (depending on how much energy was absorbed from the photon).

3. The varying calculations of Threshold Percentages during the process of Successive EM bursts.

Options of Emitting Multiple Frequencies

Up to this point we have been discussing the situation where an orbital system of the molecular bond gets excited (acquires additional energy), reaches the threshold percentage, and then emits an EM burst (while the remaining orbital system drops to a lower energy existence). However, many molecular bonds are capable of emitting one of several frequencies of EM bursts.

Scientists have known about this for a long time. Their basic understanding will be presented first. After that, we will add the new concept of threshold energies to explain which EM burst will be emitted.

Advanced Understanding using Energy Percentage

Every molecular bond has a specific set of options for the frequencies of EM burst will may be emitted. Of those specific options, which frequency will be emitted at a particular time depends on two factors:
1. The amount of energy absorbed
2. The Energy Percentage diverted to the energy field

For example, we will look at a typical molecular bond. This molecular bond can only exist at a few energy states. Stated another way, the orbital system of this molecular bond can only have certain amounts of energies and no others. In this example, there are four options for energy states.

Level 4 ———
Level 3 ———
Level 2 ———

Level 1 ———

Now we add energy to the orbital system. For best illustration of the concepts, let us add maximum energy to the orbital system, which will then make it exist with the maximum possible energy.

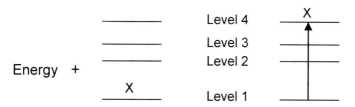

When the orbital system "decides" to emit an EM burst and exist at a lower energy state, there are three options to choose from:

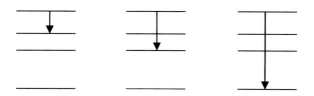

Which energy of EM burst will be emitted (of the possible options) will depend on the energy percentage diverted to the energy fields at the time of emission. For example:

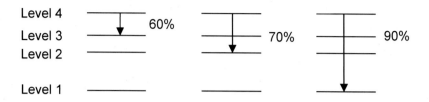

% of total energy diverted to energy field to emit each amount of energy in EM burst

In this example, to emit the first EM burst, the energy diverted to the EM field must be at least 60% of the total energy of the system. This will divert the amount of energy equal to energy of Level 4 minus the energy of Level 3. That burst will be emitted, and the orbital system will now exist at energy level 3.

In order to emit the second burst, from Level 4 to Level 2, the energy must be at least 70% of the total energy of the system. This will divert the amount of energy equal to the energy of Level 4 minus the energy of Level 2. That burst will be emitted, and the orbital system will now exist at energy level 2.

In order to emit the third burst, from Level 4 to Level 1, the energy must be at least 90% of the total energy of the system. This will divert the amount of energy equal to energy of Level 4 minus the energy of Level 1. That burst will be emitted, and the orbital system will now exist at energy level 1.

Remember that it is not just the amount of energy we start with in our orbital system, it also the percentage of that energy diverted to the energy fields at the instant the EM burst takes flight. The two factors together will determine what energy (and hence what frequency) of EM burst will be created.

> In terms of Threshold Percentage, which energy
> (and hence frequency) of EM burst will be emitted
> (of the possible options for a given molecule)
> will depend on two factors:
>
> 1. The amount of energy in the system to begin with
>
> 2. The energy percentage diverted to the energy fields at the time the EM burst is created.

Starting from a Different Level of Energy

We can start from a different level of energy, depending on how much energy we put into the system. For example, we can put enough energy to move the system to energy level 3 rather than energy level 4. From that point, the basic process of EM emission is the same. Furthermore, the options for total energies of the orbital system are the same, and the options for energies of EM bursts are the same.

The concepts of threshold energy still apply. However, the percentages required are different. Because we are starting at a different energy level, the percentage required for EM creation will be calculated according to the new orbital system total. The percentage will always be based on where the energy level starts, as compared to where the energy level is going.

Remember we have several factors: a) the total energy level of the orbital system where we start, and b) the amount of energy needed to be released to arrive at each successive lower level. These two factors determine the percentage of energy which is required to be diverted to the energy fields in order to emit each burst of EM energy.

For example, in our molecular bond when we start at energy level four, and want to emit EM burst which will result in our orbit at energy level 1, then the energy percentage required to emit is 90%. This means that 90% of the energy in our orbital system *when starting at energy level 4* must be diverted to the energy fields at the time of lift-off in order to emit that particular burst of EM energy.

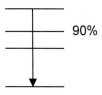

However, if we start at energy level 3, and want to reach that same energy of level 1, then the % is different. For example the required percentage to be diverted to the energy fields *may be 35% of level 3*, in order to emit EM burst which will lower energy of system to level 1.

In both examples we are using the same general concept of energy percentages diverted to energy fields. In both examples we want to reach the same lower energy level. However, our orbital systems start with different energy levels, which will ultimately change the percentage of energy needed to be diverted from the starting energy state.

Steps of Sequential Bursts and Percent Energies

The energy of the orbital system can also drop down in sequential manner. In this process, the molecular bond first emits one burst of EM energy, and the remaining orbital system exists at a lower energy. Then, from that lower energy level, a second EM burst is emitted, and the remaining orbital system lowers again to yet another level.

Taking the concepts above and applying here, we can determine the required energy percentages from each successive starting point, to reach each successive ending point. Remember, it is all related to the energy level where we start, as compared to the energy level where we wish to go.

In order to emit the desired EM burst (and also for the remaining orbital system to reach the desired lower energy level), for each successive step, we must divert enough energy to the energy fields in order to reach the required percentage – to not only create an EM burst, but to create an EM burst of the desired size.

For example:

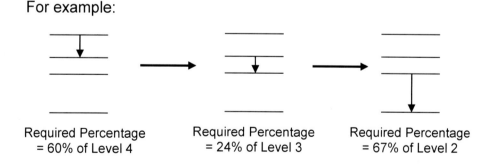

Required Percentage = 60% of Level 4   Required Percentage = 24% of Level 3   Required Percentage = 67% of Level 2

Chapter 10
Electrons as Particles and Waves

Introduction

Electrons are similar to EM bursts in several ways. One similarity is that electrons have been observed to behave as both particles and waves. We can now understand why electrons appear to have both properties.

In previous chapters we have discussed in detail why EM bursts act as both particles and waves. We can apply this understanding to electrons behaving as both particles and waves.

Similarly, in earlier chapters we discussed how electrons in molecular orbitals are able to create EM bursts from the "molecular bond".

We can take both of these concepts and apply them to electrons. We will then be able to see the one physical nature of the electron can be seen as both particle and wave.

Furthermore, in this chapter we will discuss my "General Principle of Particle-Wave Duality".

Two Types of Electrons

Introduction

It is important to know that there are two types of electrons:
1. Atomic electrons (electrons traveling in orbitals)
2. Free electrons (electrons traveling through the air)

The two types of electrons operate differently. The two types of electrons behave differently. Their wave patterns are created differently. Therefore we will discuss the operations of each type of electron separately.

Also note that some properties associated with "electron as wave" apply to electrons in the orbital, whereas other properties associated with the "electron as wave" apply only to free electrons.

At this time we will provide a brief overview of each type of electron. Further details will be found throughout this chapter, and throughout this book.

1. Atomic Electrons (Electrons in Orbitals)

The first type of electrons I will refer to as "atomic electrons". These atomic electrons are attached to the atom, held in place by a type of gravitational pull from the nucleus.

The atomic electron travels in a complex path around the nucleus. This complex path is what creates the orbital.

The "wave" of the atomic electron is essentially created by the electrical field of the electron. This field cycles in predictable patterns as the electron travels around the orbital. The net result is a "wave" of cycling electrical fields.

The energy fields and the wave patterns of the atomic electrons are important for the following processes: emission of EM bursts; absorption of EM bursts; and standing waves in orbiting electrons.

Note that most of this chapter will be devoted to the fields and wave patterns of atomic electrons. (Free electrons will become important only when we discuss interference of electron waves).

2. Free Electrons

Free electrons are those electrons which are not attached to any atom. These free electrons travel through the air like a baseball.

Note that most free electrons are artificially created. Scientists create a "beam" of free electrons. The net result is a continuous stream of free electrons, emitted from a device, much like bullets from a machine gun. This beam of free electrons is then sent into various materials to perform a variety of experiments.

The properties of the free electron are very similar to the properties of photons. Both are particles, sent through the air like a baseball. Also, both have a pulsating motion. As with the photon, when the two motions of the free electron are combined (forward motion and oscillating motion), the free electron can be tracked as a "wave". Notice that this mechanism for the wave pattern does *not occur* in the atomic electron.

The wave pattern of the free electron is most important when we discuss the interference patterns created by two or more wave patterns of free electrons. In fact, the electron was determined to be a wave because of experiments which showed electron interference patterns.

However, again, this experiment only applies to free electrons, not electrons in orbitals. We cannot assume that electrons behave in structured orbitals as they do when freely flying through the air.

Also note that this mechanism for creation of the wave pattern leads us into my "General Principle of Particle Wave Duality"

Atomic Electron as Particle and Wave: Overview

Like the EM burst, the electron itself is a particle not a wave. Whether the electron is an atomic electron or a free electron, the electron is primarily a particle. Specifically, the electron it is a collection of smaller physical objects (energy strings) which together, create an electron. (Note that new models for the electron will be presented in a later chapter).

The atomic electron does *not* travel in a wave. The atomic electron travels in a path – it may be a complex path but it is a path. During the travel along the path, the atomic electron does NOT move up and down, nor does it move side to side. It is important to know that the atomic electron itself does NOT move as a wave.

Furthermore, and similar to the EM burst, the wave pattern of the atomic electron exists as we track the motion of the energy field. Recall that in the EM burst pulsating through the air, we see the electric field pulsate up and down, and the magnetic field pulsate left and right. Then as the EM burst pulsates and moves forward simultaneously, the EM burst creates a wave pattern.

Similarly, when we see the atomic electron as a wave, what we are actually seeing is the up and down motion of the electric field, combined with the forward motion of the electron. Specifically, we are observing the cyclic nature of the electric field as the electron movies along its path. This is the wave pattern we observe in the electron as it travels in the orbital.

Electron as Particle and Wave on Power Lines
(Diagrams and Explanations)

Overview

In the following sections we will show, through diagrams and explanations, how the electron is both a particle and a wave.

The wave pattern created by an electron in an orbital is almost identical to how electrons create wave patterns as alternating current in power lines. Therefore on this first section we will look at electrons on a power line, and see how electric fields are created, which can then become wave patterns of cyclic energy fields. In the next section we will extend these concepts to electrons which travel in an orbital.

To begin, we will repeat some concepts we discussed earlier. Then we will add additional concepts, until we can finally see the electron as a wave.

Electrons and Energy Fields in Two Directions

1. When an electron moves in one direction, such as left to right, then an energy field is created. This energy field is created by the straight line motion of the electron.

2. As long as the electron moves from left to right, the electrical energy field will extend upward. Conversely, as long as the electron moves from right to left, the electrical energy field will extend downward.

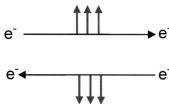

3. Note that the electrical energy fields are still attached to the electron at this time. The energy fields merely expand out as the electron moves forward. (We will not consider any threshold energies or EM lift off in this discussion of electron waves).

Electrons and Energy Fields on AC Power Lines

4. When an electron literally moves back and forth, such as on a power line of alternating current, the electrical energy fields and electron travel look like this:

4a. Left to Right motion of electron:

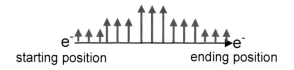
starting position   ending position

    The electron starts out with no forward motion. Then it gradually picks up speed. As the electron travels faster, the electric field will extend out further. Then, as the electron reaches its ending position (in this case, because it is being pulled back by the generator), the electron slows down. As the electron slows down, the electric field becomes smaller and smaller.

4b. Right to Left motion of an electron:

new ending position   new starting position
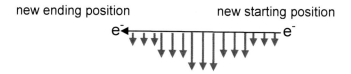

    The right to left motion of the electron is similar, but in reverse directions. The electron starts out with no forward motion. Then it gradually picks up speed. As the electron travels faster, the electric field extends out further. [This time the electric energy field is extending down rather than up].
    Then, as the electron reaches its new ending position (in this case, because the generator is decreasing its pull, and will soon start another forward push), the electron slows down. As the electron slows down, the electric field becomes smaller and smaller. [Again, the electrical field is extending down during this time when the electron is on the right to left path].

Tracking the Energy Fields on the Power Line

5. Tracking the energy fields, as created by the electron traveling on the power line, would then look like this:

Left to Right motion of electron

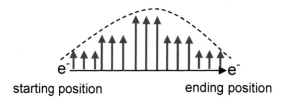

Right to Left motion of electron

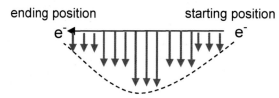

Creating the Wave Pattern

6. We can then draw these graphs in a slightly different way. Instead of looking at the how the energy field changes in relation to direction of electron travel, we look at that same "how the energy field changes" but as a function of time.

Here our X-axis would be the progression of time, rather than the direction of electrons. And yet, the tracking of the energy fields will be essentially the same. This means that the graphs above will essentially be placed side by side, like this:

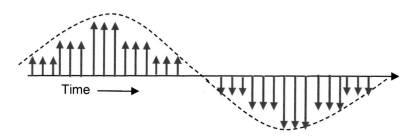

7. From this you can easily see the Wave Pattern!

The wave pattern of the electron on a wire is actually the cyclical production of electrical fields. By tracking the extending and retracting of electrical field, first in one direction, then the other, we observe a wave pattern.

Continuous Wave Pattern on a Power Line

8. Furthermore, remember that on an Alternating Current Power Line this action happens repeatedly, over and over again, as long as the generator at the power plant keeps spinning.

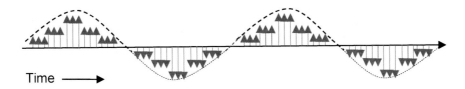

Summary of Wave Pattern on a Power Line

Therefore, electrons produce a wave pattern on a power line as follows: The generator at the power plant pushes electrons forward and pulls them backward. As the electrons travel left to right, the electrical field extends upward. As the electrons travel right to left, the electrical field extends downward.

Because the generator at the power plant continuously creates these forward and reverse motions of electrons, the electrical fields continue to extend up and down, and then back up again.

If we track the electrical energy fields over time, rather than associated with direction of electron, we can see that the cyclic electrical fields produces a wave pattern of energy over time.

In total, this cyclic process of electrons moving back and forth, produces electrical fields that extend up and down, and thereby creates a wave pattern over time. This is how we see the wave properties of an electron on a power line.

In the next sections we will extend this discussion to show the oscillating energy fields produce wave properties of electrons in atomic orbitals.

Electron as Particle and Wave in Orbitals
(Diagrams and Explanations)

Overview
  Now we can look the path of an electron around an atom and see how this electron will be both particle and wave.
  We can take everything we have learned previously for electrons traveling on a wire, being pushed and pulled, and creating the alternating electrical fields...and apply that to electrons traveling on an elliptical orbit. Doing this we can see how wave patterns are created by electrons in an orbit.
  Further, we can extend these concepts to orbits of any shape. Thus, by the end of this section you will see how an electron in any shape orbital will produce a wave pattern.

Energy Fields on an Orbital Path
9. Let us look at one electron on an orbital path. We begin with the electron traveling to the right. As the electron travels left to right, an electrical energy energy field is also created, extending upward.

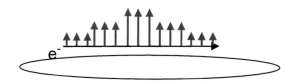

10. As the electron rounds the bend, through the atom and out the other side, the energy field changes. As the electron reaches the end of the elliptical orbit, the energy field shrinks, until it is almost non-existent.

11. Then as the electron changes course, the energy field begins to flow in the opposite direction.

When the electron rounds the bend, it proceeds to travel in the opposite direction. In our example, the electron is traveling from right to left. As the electron travels, another energy field is created. This time the energy field flows downward (exactly opposite from before).

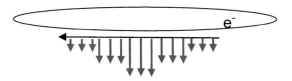

12. The electron then rounds the bend on the other side. Again, the energy field becomes smaller and smaller, to almost non-existent. Then as the electron goes to the other straight side of the ellipse, a new energy field begins to grow.

Tracking the Energy Fields on the Orbital Path

13. Tracking the energy fields, as created by the traveling of the electron around the orbital, would then look like this:

Left to Right of elliptical track of electron

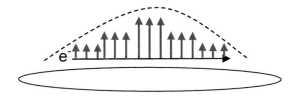

Right to Left of elliptical track of electron

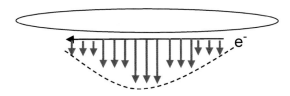

Creation of the Wave Pattern in an Atomic Orbital

14. Then we can draw these graphs as how the energy field changes as a function of time.

Here our X-axis would be the progression of time, rather than the direction of electrons. And yet, the tracking of the energy fields will be essentially the same. This means that the graphs above will essentially be placed side by side, like this:

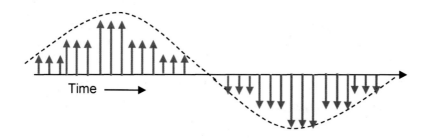

15. From this you can easily see the Wave Pattern:

The wave pattern of the electron in an orbital is actually the cyclical production of electrical fields. As the electron travels left to right, and then right to left, on each track of the elliptical orbital, the upward and downward electrical fields will be produced. Tracking the extending and retracting of the electrical field, first in one direction, then the other, is what creates the Wave Pattern we observe.

Continuous Wave Pattern in Orbital

16. Furthermore, because the electron path on an orbital is continuous – the electron keeps on this same path essentially forever – then the cyclic nature of electric fields is also continuous.

As long as the electron continues to travel through its orbit, the electron will produce a cycle of electric fields extending upward, then downward, along each side of the elliptical path. This produces a continuous upward and downward extension of electrical fields, which can be graphed over time as follows:

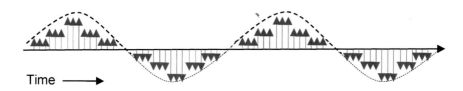

Summary of Electron Waves Created in an Orbital

Therefore, electrons produce a wave pattern on an orbital as follows: The electron is a particle which travels a continuous elliptical path. When we have an electron traveling in an elliptical orbit, we will have two types of cycles: alternating current and alternating energy fields.

As the electron travels left to right, the electrical field extends upward.

As the electron travels right to left, the electrical field extends downward.

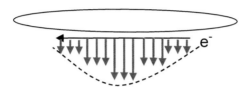

Because the electron continuously travels this path, these alternating energy fields will continue repeatedly as the electron continues to travel the loop. The electrical fields continue to extend up and down and back up again. If we track the electrical energy fields over time, we can see that the cyclic electrical fields produce a wave pattern of energy over time.

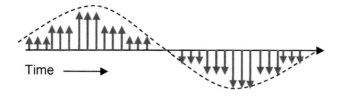

In total, this cyclic process of electrons moving back and forth produces electrical fields that extend up and down, and therefore creates a wave pattern over time. This is how we see the wave properties of an electron in an orbital.

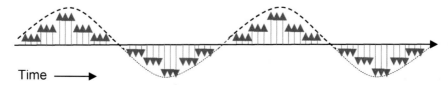

Electric Fields Always Face Nucleus

Overview

The electrical energy fields are drawn flowing up and down for simplicity. However, in reality when an electron travels around an atom the energy fields flow more horizontally than up and down.

The cause is simple: the electrical energy fields always flow toward the nucleus. In fact, it is because the electrical energy fields always flow toward the nucleus that the energy field produces the pattern described above.

Why Electric Field Always Points Toward Nucleus

Why does the electric energy field always point toward the nucleus? The details will be explained in subsequent books (including the new model of the electron and new model of gravity). Yet we can discuss a few basic concepts at this time.

In brief: the electric energy strings of the field are attracted to the nucleus to the gravitational pull of the nucleus, and aided by the opportunity to merge with the proton.

Remember that the electric energy field is actually a set of strings, and these strings have mass as well as energy. Therefore, because the strings have mass, they have gravity. Thus, the mass of the nucleus will pull not only on the electron as particle, but pull on the energy field strings as well.

Further, with a larger mass, we have a stronger gravitational pull. Therefore, with a larger nucleus each energy field string is pulled more strongly toward the nucleus.

In total, this means that no matter where the electron itself is, the energy fields will be attracted to the nucleus. The net effect is that the electrical energy strings are always being pulled toward the nucleus.

Direction of Energy Fields and Wave Pattern from Nucleus

Therefore, in reality the energy fields usually point inward toward the nucleus, rather than pointing up and down.

The basic process is the same as described above. However, instead of energy fields extending "up" and "down", imagine the energy fields pointing "inward from Side A", then "inward from Side B".

The basic effect is the same: energy fields which change direction as the electron changes direction. However, we are more accurate by saying that the energy strings change direction, as the electron changes direction, so that the energy strings always point toward the nucleus.

Straight Paths and Going Around the Bend

This also explains the directions of the energy fields illustrated above. Think of the ellipse above as surrounding the nucleus, with the energy field always pointing at the nucleus. This is what we then observe:

a. As the electron travels left to right, the energy field points inward. This energy field will remain pointed in that same direction, as long as the electron is on that straight segment.

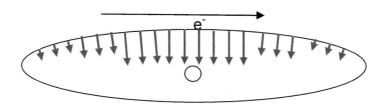

b. As the electron rounds the end of the ellipse, the energy field shrinks to zero. Why? Because the electrical energy strings are preparing to exit through the electron, and outward through the other side of the electron.

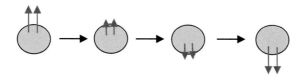

c. As the electron travels right to left, the energy field again points inward. Yet *the field does this from the other side of the electron.* (This will be more fully illustrated in the later chapter on the new model of the electron). Then, as above, the energy field will extend from this side of the electron, toward the nucleus, for as long as the electron travels this straight segment.

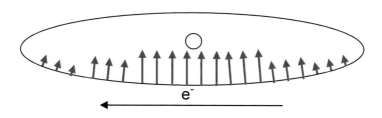

d. As the electron rounds the end of the ellipse, the energy field again shrinks to zero. Just as in step b, the electrical energy strings are preparing to exit through the electron, and outward through the other side of the electron (which is actually the original exiting side of the electron as in step "a".)

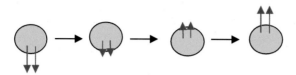

The Wave Pattern

We can of course look at the orbital from any angle, and therefore observe the wave pattern of energy fields from any angle. Therefore, the oscillating energy fields will of course produce the exact same wave pattern as before:

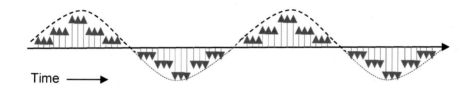

Complex Orbital Shapes and Complex Wave Patterns

There are many orbital shapes, and the orbital can be very complex, yet the basic principle is the same: As the electron changes direction, the energy field changes direction. Therefore the wave pattern is created.

Thus we can apply all of the concepts described above to any orbital. The orbital shape can be very complex, with many twists and turns, yet the electrical energy fields will extend, retract, and extend in the opposite direction depending on the particular direction the electron is traveling. All of this can be tracked as a wave of energy extending, retracting, and extending in the opposite direction, over time.

Therefore, a complex path of the electron, in a complex orbital shape, will create a complex wave pattern of electrical energy.

The principle is always the same, even though the resulting energy field wave pattern may be quite complex.

The Explanation for the Electron as Both Particle and Wave in an Atomic Orbital is as Follows:

The electron is first and foremost a particle. However, the electron in motion creates oscillating electrical fields. The pattern of the oscillating electrical fields, as the electron travels the orbital, can be tracked as a wave.

1. When the electron travels in an orbit, either an atomic orbit or a molecular orbit, the electrical energy field changes direction.

2. As the electron travels to the right, the electrical field flows upward. Conversely, as the electron travels to the left, the electrical field flows downward. This creates a wave pattern of oscillating electrical energy.

3. Because the orbit is a closed loop, the electron repeats the process over and over again. This creates a repeated wave pattern of electrical fields. Therefore, along with the particle itself traveling around the orbit, we also have a wave pattern of energy fields.

4. Therefore, because of this situation, we have the atomic electron as a particle, with the oscillating electric fields as a wave. The atomic electron will then be seen as both particle and wave as the electron repeatedly travels its closed loop path.

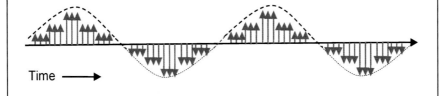

Molecular Bond and Fields Pointing Toward Nuclei

In a molecular bond, the situation is similar. However, the fields point to one nucleus, and then to the other nucleus. This is because the electric fields themselves are attracted to one nucleus, and then the other, as the electron itself gets close to each nucleus.

Notice that this will also produce a more sophisticated type of wave pattern.

In brief: we first have the field changing directions to face nucleus A, as in diagram above. But then as the electron moves closer to nucleus B, the field changes its attraction, toward nucleus B. Then as the electron goes around nucleus B we have a similar process of changing fields toward nucleus B as we did for nucleus A. Thus, we essentially have two oscillating field wave patterns, one for each nucleus.

Yet these patterns are repeated very regularly as the electron travels its molecular orbital path. Therefore, we have regular and predictable parameters of the oscillating energy fields, including the standing wave.

(This will be explored and illustrated in greater detail in the book "New Models of Electrons, Orbitals, and Atoms").

Standing Waves of Electrons

Overview

A "standing wave" is any wave which appears to be constant, and is confined to a particular region. For the electron wave of an orbital, the wave appears as a closed wave, where the end is the same as the beginning. Similarly, this wave appears to be confined to one specific region (thus the wave is referred to as "stationary" or "standing").

Requirements for Standing Wave

A standing wave will only exist if the following conditions are met:
1. The wave is repeated continuously
2. The wave is repeated in a specific region of space, and
3. The wave is the exact same wave pattern, no matter how complex that pattern is.

Creation of Standing Wave by Electron

The electron which travels in an orbit can create a standing wave. Now that we understand how the electron creates a wave when traveling in an orbital, we can understand the existence of the standing wave in an orbital.

The electron travels around the orbital in a loop. The path may be complex, but the end of the path is always the beginning…and the electron repeats the path.

The electrical fields created by the electron will then also be repeated. The electrical fields which extend from the electron will extend in the same direction, to the same amount, at the same region of physical space. As the electron repeats its path, the energy fields will also increase, decrease, and change direction as happened the previous time the electron completed the path.

Stated another way, the wave pattern created by the fluctuation in electrical energy field will be the same every time as the electron repeats its path.

This is the creation of the standing wave as created by an electron in an orbital.

Complex Standing Waves

Note that these "standing waves" can be extremely complex. As long as the three requirements above are met, the electron can create a standing wave.

Again, those requirements are: a) if the wave is repeated continuously, b) the wave is repeated in a specific region of space, and c) the wave is the exact same wave pattern, no matter how complex that pattern is.

For example, imagine a very complex race track. There are numerous twists and turns. In fact, the path also goes up hills and down hills, as well as making sharp turns. And of course there are some segments of the path which are straight.

However, this is the main thing: the path is a closed loop. No matter how complex the race track is, the car will repeat the exact same path again (and again and again). Consequently, the electrical field will increase and decrease, in the same amounts, and at the exact same regions again and again. The net result is not only a wave pattern of rising and falling of electric fields, but the exact same pattern of rising and falling electric fields. Hence: a standing wave.

Therefore, any electron in an orbital, no matter how complex that orbital shape is, will be capable of producing a standing wave…a repeating pattern of oscillating electric fields…as long as those three requirements above are satisfied.

Electron as Blurred Energy

Overview

Electrons are sometimes described as being a blurred object, a fuzzy entity zipping through space. Some scientists refer to this as a smeared wave. Now we can understand why this is so. Essentially, the electron is traveling so fast that the wave pattern of electrical energy appears to us as a blur.

Blurred Standing Wave

Above we explained how:
a. Electrons produce electrical energy fields
b. These energy fields will be created in a wave like pattern as the electron travels in the orbital, and
c. A standing wave occurs as the electron repeats its path, no matter how complex.

If the electron traveled slowly enough, we might be able to see the full path, and be able to see the full energy field wave pattern. However, the electron travels much too fast for us to see. Therefore, all we get is a blur.

This is similar to the hummingbird. The wings are moving up and down repeatedly. Yet the wings move so fast, that our eyes can see only a blur of the undulating wings.

The electron is very similar. The wave pattern is so fast, that all we see is a blur.

Electrons in Complex Patterns and Limits of Our Measuring Devices

Furthermore, electrons around an atom travel in a complex pattern. This produces the orbital. More specifically, the complex path of the electron and its complex energy wave produce the complex shape of where an electron might be at any one time. Some people view this as another type of blurry electron.

In reality the energy field is upward at one location, then downward at another location, then upward again at a third location...as the electron travels through its orbital path. These are actually distinct locations of energy fields. Yet these oscillating energy fields occur so fast that our measuring devices will pick up all these energy fields, of all these positions, at the same time. The net effect, from the measuring device or observer, is in fact a blurred image, as if the electron is everywhere at once.

This is much like the long-exposure photography of cars on the highway. The resulting picture is a blur of the car's path. A similar effect occurs with our measuring device and the complex path of our electron. The electron is not really in multiple places at once, any more than the car is at multiple locations at once. Rather, it is our observation device in relation to the speed of the electron which captures all the positions of the electron and all the energy fields produced, as all being simultaneous. The net effect is a blurred image of the electron and its energy.

Furthermore, because we have a standing wave – that is, the energy field waves occur in a complex pattern that repeats forever – that our snapshot will always give the illusion of blurred energy.

Smeared Electron as an Illusion

Thus the electron appears to be "smeared", but in fact this is an illusion. The electron is in distinct locations at different times, and the energy fields are produced at different times, yet the speed of the process, and the existence of the repeating pattern, gives us the illusion of a smeared object. The electron appears to be everywhere at once, and the energy waves appear to be everywhere at once, when in fact this is not the actual process.

Again, it is due to the speed of the electron in relation to our observation devices, along with the standing wave pattern, that we see a smeared object - a blur of energy as if everything happened simultaneously.

Creating the Electron Wave for a Free Particle

Overview

There is another process which can create the wave pattern for the electron. This process involves the motion of the electron itself, rather than the electric fields, to produce a wave pattern. Note that this mechanism is observed only in free electrons, not in electrons attached to the atoms.

Electron Vibration and Forward Motion

The basic mechanism of this process involves two steps: vibration and forward motion. Together, these simultaneous motions produce a wave pattern. This process is similar in many ways to the combined motion of the EM burst.

Note that this process occurs primarily for free electrons flying through space. Also note, as we shall see later, that this process can apply to many particles (as long as each of those particles satisfies certain criteria).

Electron Vibration

An electron has numerous motions, one of which is vibration. What is vibration? Simply put, the vibration of an electron is small movement in one direction, then a small movement in the other direction, at a very fast rate.

Electron Vibration

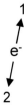

Forward Motion

The electron also travels in a forward motion. As long as the electron is moving "forward" in some way, it qualifies. For simplicity, we will first consider the free electron flying through space.

Forward Motion

e⁻ ───────────────────────►

Combining Vibration and Forward Motion for Wave Pattern

When we combine the vibration motion of the electron and the forward motion of the electron, we will see a wave pattern. The combination of these two motions will produce a total motion which can be viewed as a wave.

Combined Motion of Vibration and Forward Travel

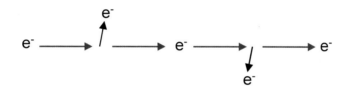

Wave Pattern of Combined Motion

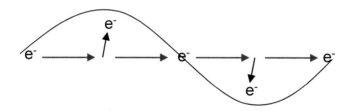

Thus, the motion of the electron particle can be tracked as a wave!

The Alternate Process:
How a Free Electron can be Seen as a Wave:

1. The electron vibrates; this vibration is actually small motions in one direction and then another.

2. This vibration motion, combined with the forward trajectory of the electron, produces a total motion of the electron particle which can be tracked as a wave.

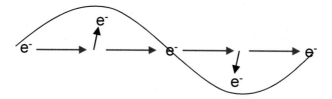

3. This effect may be produced in both the free electron and the atomic electron. However, the effect is much more significant in the free electron.
   The lack of forces from the nucleus allows the free electron to move further up-down, and thus the effect is more noticeable.

Two Methods of Electron as Particle and Wave Reviewed
(Electrical Energy Field versus Electron Motion)

Overview

Based on earlier discussions, we now know that there are in fact two mechanisms which can make the electron particle appear to be a wave: oscillating energy fields, or the vibration of the electron combined with its forward motion.

We must first remember that he electron is first and foremost a particle. The electron is a particle and will always remain a particle.

What makes the particle appear to be a wave is the motion of the particle. In an orbital, the electron will have oscillating energy fields which produce a continuous wave of blurred energy. As a free electron, the vibration of the electron along with the forward motion will produce a wave motion.

Oscillating Electrical Field as Wave Pattern in Orbital

The first mechanism is the electric field. As the electron travels back and forth along the wire, or travels in its complex orbital path, the electric field extends up and down accordingly. The strength and direction of this electrical energy field can be tracked as a wave. This is the primary cause of electron as wave that we observe in atomic and molecular orbitals.

Combined Particle Motions as Wave Pattern in Free Electron

The second mechanism is the motion of the electron particle itself. The electron has many motions, including vibration and forward motion. Combining the vibration and forward motion, the net motion of the electron can be tracked as a wave. This is the primary cause of electron as wave that we observe in the free electron.

Why Not Other Mechanism in Other System

If there are two mechanisms for the creating the electron as wave, then might both mechanisms be observed in both situations? Theoretically yes, but realistically no.

We will explore some of the possibilities below.

1. In the free electron, the energy field does *not* oscillate, simply because there is no nucleus. The energy fields have no reason to aim one direction over another. Therefore in a free electron the energy fields essentially remain fixed in place while the electron itself vibrates and rotates.

2. However, the energy fields do "change direction" in the sense that as the electron spins, the fixed energy fields also rotate. (Think of a needle on a baseball: as the baseball spins, the needle spins as well). This will produce some change in energy fields, and may in fact produce an oscillating pattern, yet not the same process, or the same degree, as in the atomic orbital.

Also, in this case the actions of the electron motions are usually more of interest than the energy fields, and therefore we focus more on the wave pattern from motion of the electron rather than the wave pattern from the energy fields.

3. In a free electron, the up and down motion of the electron is more pronounced, primarily because the free electron has more internal energy.

In order to become a free electron this electron must have enough energy to break away from the gravitational pull of the nucleus. Therefore, most free electrons will have more internal energy than most atomic electrons (depending which atoms to which we are comparing the free electron).

Therefore, the free electron has more internal energy. This means it will vibrate up and down to a larger extent. (It will also possibly travel forward at a faster rate). If the electron vibrates up and down to a larger extent, then the wave pattern becomes much more noticeable.

4. In the atomic electron, the electron does *not* move up and down as much as the free electron does because of: the internal energy of the electron, and the gravitational pull of the nucleus.

Restated from above (in the opposite way) the internal energy of the atomic electron is usually less than the internal energy of the free electron. This is because the atomic electron does not yet have enough energy to break free from the nucleus.

And since the atomic electron does not have as much internal energy, the vibrational motion will be less. Therefore, the up-down motion of the atomic electron will be much less significant, if it exists at all.

5. Additionally, the gravitational pull of the nucleus is pulling on the atomic electron itself. Therefore, any internal vibration that does exist in the electron is immediately counteracted by the gravitational pull from the outside. Thus, again, the vibration of the atomic electron, though it does exist, will not be large enough to produce a significant wave pattern.

Conversely, in the free electron there is no nucleus, therefore no gravitational pull, and therefore the vibration that is produced is allowed to be as full as it can be. Hence, in the free electron we will get the full vibration portion of the particle's wave pattern.

Thus Two Mechanisms Possible, Depending on Circumstances

These are two different mechanisms, yet both produce wave patterns. Therefore, when scientists are using equipment which senses the electron as a wave, the underlying cause will be one or both of these two mechanisms.

It is important to understand the differences between the two processes so that we do not improperly assign characteristics to a particle or a situation which do not exist.

General Principles for Particle-Wave Duality

Overview

We can extend the above discussion to a more general principle on particle-wave duality.

The general principle for particle-wave duality is as follows: any particle can be seen to be both a particle and a wave if that particle oscillates and moves forward at the same time. Thus, we start with a physical entity, which produces a wave-like pattern of motion.

However, as we have seen above, there are two mechanisms for producing wave patterns: oscillating energy fields (which occur mostly in structured systems like atoms) or vibrating particles (which occur when the particle is free, having no external forces acting upon it.

Therefore, our general principle of particle-wave duality will actually be developed into two General Principles of Particle-Wave Duality.

The Free Particle

We will start with the Free Particle. Why? Because all experiments which "show" that a particular particle has wave-like properties are always using the Free Particle, not the particle bound in the atom. Thus, the original "observed" waves will be due to the principle based on the free particle.

Caveats Regarding This type of Particle-Wave Duality

As I have said numerous times, it is a mistake to assume that a free particle behaves the same way when it is in a structured system like an atom. Therefore, the principle discussed here applies only to free particles.

It is also a mistake to assume that all free particles have wave-like properties. Subatomic particles will travel as a wave, but not all objects in general; and this will be better understood when we discuss the structure of electrons and other subatomic particles in future books.

Therefore, we must remember that ONLY if the particle has certain criteria will it behave as a wave and particle.

Specifically, we can see that any subatomic particle can be tracked as a wave if that particle vibrates and moves forward at the same time.

Example: Proton

For example, consider a free proton flying through space. Suppose that this proton also vibrates. The net result would be the following combination of motions:

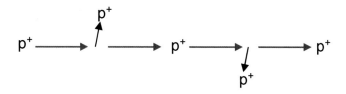

This will produce a wave pattern for the proton, just as it did for the electron. Again, the combination of motions would produce a total set of motions which can be tracked as a wave.

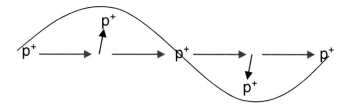

Energy Field Not Required

Note that regardless of whether the particle produces an energy field, the particle can be seen as a wave (tracked as a wave) because of this process. Therefore, any particle has the potential to move in a pattern which can be tracked as a wave.

Therefore, the First General Principle of Particle-Wave Duality is as follows:

> ## General Principle of Particle-Wave Duality #1:
>
> Any particle can be tracked as a wave regardless of whether or not it has an energy field if the particle:
>
> 1. Is a Free Particle
>   (such as flying through the air)
>
> 2. Oscillates Up and Down Repeatedly
>   (pulsates, vibrates, or migrates;
>   in one direction and then the other, repeatedly)
>
> AND
>
> 3. Travels Forward at the Same Time.
>
> This combination of motions will produce a wave pattern, for any particle; thereby creating Particle-Wave Duality

The Atomic Electron and Other Bound Particles

This brings us to our Second General Principle of Particle-Wave Duality. This principle is similar to above, however we are using the oscillating energy fields, rather than the vibration of the particle, for our up and down motion.

As illustrated earlier, this is the process by which the electron in an atom or molecule will seem to exist as a standing wave or blurred wave. Specifically: the particle moves in a repeated loop, while the energy fields oscillate. This produces a regular pattern of field oscillations while the particle moves forward. Thus, the blurred, standing wave of energy field is created.

Also note that if the particle has oscillating energy fields in other circumstances, while moving forward, then this principle will also apply.

Therefore, the Second General Principle of Particle-Wave Duality is as follows:

General Principle of Particle-Wave Duality #2:

Any particle can also be tracked as a wave, if it:

1. Has an oscillating *energy field*

    AND

2a. The particle travels in a continuous loop (such as an orbital)

    OR

2b. The particle travels in a forward motion through the air

This combination of factors will produce an oscillating energy field, which repeats continuously, and can then be tracked as a wave.

Summary of All Concepts Discussed in this Book, Focusing on New Discoveries and Models

Overview

The following are ALL the major concepts presented in this book. New theories and new models which are created by the author will be highlighted with the following: (new)

More specifically, ALL concepts and models discussed in this book will be listed here, with notes as I believe to be important. All concepts, models, and analogies will be listed by chapter.

Yet, the vast majority of concepts presented in this book are new concepts. The majority of concepts are in fact new discoveries. Each new discovery will be highlighted.

Purposes of this List

There are several purposes of this list:

1. The reader can easily see, at a glance, each concept. The reader can refer to this list and see the main concepts presented throughout the book.

This is the main purpose of the Summary Pages. It is an aid for the reader, to refer to these pages, to read and re-read the summaries, in order to get a better understanding of all concepts presented in this book.

2. There are many new discoveries presented in this book. Therefore the reader can refer to this list. Because many concepts are new to the world they are only presented here, and therefore should be understood as easily as possible. Again, as an educational aid, the reader can see the summary points of all concepts.

3. Furthermore, the reader can see how these many new concepts are interrelated. As the reader flips through the pages and reads various concepts, he can see how these concepts are interrelated. This will give the reader a much broader and much more subtle understanding of everything discussed in the book.

Also, by flipping through the shorter summaries the reader can see how the new concepts and new models explain multiple processes and multiple observed phenomenon. Overall, the reader will gain a much better sense of how these new concepts must be true…because they all fit together in such a tidy and comprehensive way.

4. Of course I want personal credit for all new discoveries. I also want credit for new models, new theories, new analogies, and new illustrations. Therefore I want to emphasize to each reader which of those discoveries and models are mine.

By highlighting those new discoveries and models, readers (particularly those in the scientific community) can appreciate what I have to offer, and will acknowledge me for my contributions.

<u>Organization of the Summary List</u>

These summary points are organized chapter by chapter. Then, for chapters which are longer or which have more information, the summary points are further subdivided by general topic.

When the summary point is a new concept, the word "new" will be listed after the summary point.

<u>Significance of my Discoveries, and Credit for my Contributions</u>

In addition, many of these new concepts will be further elaborated on, in terms of what makes the concept new, or why the concept is a significant advancement in our understanding of science.

These additional elaborations are mostly written for readers who do not have a detailed background in quantum science, so that they can appreciate the significance of the discoveries.

Also, some of the concepts will seem so "obvious" when explained clearly that it will be too easy for others to take the credit for my work. Also, some scientists might say "I already knew that" when in fact they did not.

Therefore, while it is true I mostly want to present these discoveries for the benefit of society, I also want proper credit. I am simply elaborating with enough details to ensure that my contributions are indeed credited to me.

Chapter 7: Creation of EM Bursts
from Threshold Percentage and Driver Strings
Summary of Concepts

General Mechanism for Creation of EM Bursts

84. Most bursts of electromagnetic energy are created using the same general mechanism. This mechanism requires:
    a. Alternating electrical current, and
    b. A high enough energy to meet the threshold for emission.

(New)
*Scientists and engineers have known about requiring alternating current and enough energy. However, the "Threshold for emission" is a totally new concept. Also, this mechanism applies to all EM bursts that I have studied. Therefore, this statement is New.

Structure of the Electron System

85. The electron is essentially an object like a planet. Attached to this electron are energy strings – both electrical energy strings and magnetic energy strings. The total structure of the electron and all of its energy strings I refer to as "the electron system". (New)
    *This is a new model of the electron. A major new concept

86. The energy strings in the electron system are like Legos: these energy strings can be built up, taken apart, and rearranged.
    (New)
    *This is a major new concept.

87. Whenever new energy is added to the electron system, some of the energy will go into the arrangement of the energy strings, and the rest of the energy will go into the electron motion.
    (New)
    *This is a major new concept.

88. When energy is added to the electron system, and that energy is put into the energy strings, this can mean any of the following:
    a. More energy strings created
    b. Energy strings added to existing energy strings
        1. Thicker energy strings
        2. Longer energy strings
    c. Different arrangement of energy strings within the system

(New)
*This is a new understanding to the world.

89. When energy is added to the electron system, and that energy is put into the motion, that energy can go into any of the following:
    a. Increased forward speed
    b. Increased vibration
    c. Change in spin

Note that the change in orbital (region of space for electron) is associated with a significant change in speed, as is discussed in a later chapter.

Energy Percentage

89. The "Energy Percentage" is the amount of energy diverted to the energy fields (rather than electron motion) as compared to the total energy of the system.
    This concept of the Energy Percentage is significant when understanding the process for emission of electromagnetic energy.
    For example, the difference between a power line and a transmission antenna is primarily the Energy Percentage in each. The power line will have a lower energy percentage – resulting in more electron motion, and less emission of EM bursts. In contrast, the transmission antenna has a much higher energy percentage, where most of the energy is used to create EM bursts, and very little into forward electron motion past a certain point.
    (New)
    *This is a very new concept, and a very big deal.

Energy Strings Attached and Breaking Free

90. Energy strings (both electric energy strings and magnetic energy strings) remain attached to the electron most of the time.
   (New)
   *This is a very new concept.

91. When an energy string has sufficient energy, it will break free from the electron, and become an independent entity floating through the air.
   (New)
   *This is a very new concept.

91. A burst of electromagnetic energy is in fact composed of electric energy strings and magnetic energy strings. These energy strings come from the electron. Therefore when a group of electric energy strings and energy strings manages to break free from the electron at the same moment, a burst of electromagnetic energy is created.
   (New)
   *This is a very new concept. And very important.

92. In order for a burst of electromagnetic energy to be launched from an electron, the group of energy strings (which will become the EM burst) must have enough energy to launch from the electron. This is similar to a rocket having enough energy to launch from the earth into space.
   (New)
   *This is a very new concept. And very important.

Threshold Percentage

93. Furthermore, launching the EM burst is not just a specific amount of energy, but is in fact a percentage of energy relative to the whole system. In other words, any EM burst launched from the electron will have certain percentage of energy in the energy strings relative to the total energy. (The total energy is the energy in the motions + the energy in the strings).
   (New)
   *This is a very new concept. And it will be more significant when we talk about different frequencies of EM bursts being emitted.

94. The "Threshold Percentage" or "Energy Percentage Threshold" is the required amount of energy to be diverted to the energy strings (rather than electron motion) as compared to the total energy of the system, in order to launch a burst of electromagnetic energy.

(New)

*This is a very new concept. I came up with the term "Threshold Percentage" to describe the amount of energy required for strings to break free as an EM burst, and yet *also* take into account the option of multiple possible frequencies of EM bursts from the same electron system.

95. The burst of electromagnetic energy is emitted in two stages. First the EM burst is built, then the EM burst is launched. This is very similar to a rocket being built, then the rocket being launched.

In practical terms, this means that first the ideal arrangement of energy strings is created. This means the proper number of strings and the proper thickness. This arrangement is what will be launched as independent EM burst.

Then there must be an additional amount of energy to actually launch the EM burst into the air. Thus, for stage two we begin with what will be the EM burst that will exist when launched. Then we add just enough energy to the energy strings for those existing energy strings to separate from the electron. At that point, we have launched the EM burst.

(New)

*This is a very new concept, and a big idea in understanding the actual emission process of the EM burst.

Internal Energy and Launch Energy

96. There are actually two types of energies involved in the launching of the EM burst. (This is aside from electric and magnetic). I refer to these to energies as the "Inherent Energy" and the "Launch Energy".

The "Inherent Energy" of the EM burst is the energy of the burst when it is an independent entity. It is what we know of as the energy of the photon. And, as stated above, it is the total energy of all the energy strings in the EM burst.

The "Launch Energy" is the energy required to actually launch the EM burst off the electron. Note that the launch energy must be added to the system of electron strings, and yet this energy does not become contained within the independent EM burst. The launch energy is similar to a push, where an outside force gets the object started in its motion.

(New)
*This is a very new concept, and helps understand the actual emission process of the EM burst.

97. Note that the "Threshold Energy" described earlier is actually a combination of the inherent energy and the launch energy.

(New)
*This will be important in later chapters.

98. The "Inherent Energy" and the "Launch Energy" are also responsible for the two main types of EM motions (pulsation and forward motion).

As stated above, the "Inherent Energy" of the EM burst is the total energy of the energy strings in the EM burst. Therefore, the Inherent Energy is solely responsible for the pulsation frequency of the EM burst.

Also as stated above, the "Launch Energy" of the EM burst is the applied additional energy which is required to launch the EM burst. Therefore, the Launch Energy is solely responsible for the velocity of the forward trajectory of the EM burst.

Stated another way: the Inherent Energy and the Launch Energy are the two types of energy which are initially responsible for the dual nature of the EM burst. The Inherent Energy supplies the energy for the frequency of pulsation, while the Launch Energy supplies the energy for the forward motion.

(New)
*This is a totally new insight which I have made.

Energies and Motions as Related to Mass

99. Both types of energy (inherent energy for frequency of pulsation, and launch energy for the forward trajectory energy) are related to the overall mass of the strings. The explanations and reasoning is slightly different for each, and yet both are ultimately related to the mass of the energy strings.
  (New)
  *This is a deeper insight into the understanding of electromagnetic energy. This shows that I have made correlations and looked at things from numerous viewpoints.

100. Remember that the "Inherent Energy" is the total energy of the energy strings which will be in our EM burst when launched. Whether launched or attached to the electron prior to launch, this group of energy strings is the "Inherent Energy" of the EM Burst.
  More specifically: as stated above, the arrangement of energy strings must be created before it is launched. This is similar to a space craft being built before it is launched. Therefore, all the strings are arranged in a particular way, with a particular thickness, prior to being launched. This group of energy strings will become the energy strings within the EM burst when it is launched. Therefore, the internal energy is in fact this arrangement of energy strings, whether it is in the air as an independent EM burst, or whether it has not yet left the electron.
  (Partially New: Mostly Review of earlier new concepts, and stated in a different way)

101. The inherent energy (which will create the frequency of pulsation) is related to the mass of the strings. In this situation, the inherent energy exists first, within the total energy of the energy strings. Yet because the energy strings have mass, the inherent mass of the EM burst is directly related to the inherent energy of that EM burst. (New)

102. The frequency of pulsation is caused by the inherent energy, and more specifically caused by the mass of the strings. (New)
  The cause of pulsation is related to both the mass and the energy of the strings (this is explained in a later chapter), and the specific frequency of pulsation is related to the specific mass of the strings. Therefore, the frequency of the pulsation is directly related to the mass of the strings.

103. Remember that the Launch Energy is the amount of energy required to separate a group of energy strings from the electron.
(Partially New: Review of New Concept Above)

104. The Launch Energy is the same as the Kinetic Energy of Forward Motion.

For the forward trajectory of the EM burst we can view the EM burst as a simple particle like a baseball. In general, the kinetic energy of any forward moving particle is essentially the same as the external energy applied to that particle to get it moving. In other words, for our EM burst the specific amount of Launch Energy required to lift the group of energy strings away from the electron...is also the amount of energy the EM burst will have as it travels in its forward direction.

Therefore, the Launch Energy and the Kinetic Energy of Forward Motion are the same amounts.
(Partially New)

105. The Launch Energy required to launch a burst of electromagnetic energy is also related to the mass of the EM burst to be launched.

a. We can first look at this is in a simple way, such as a space vehicle. With a larger mass space craft, we will require more energy to launch that craft from the earth into space. Similarly, with a larger mass EM burst, we will require more energy to launch that EM burst from the electron into the air.

b. We can also understand this by looking at the kinetic energy. The forward trajectory of the EM burst is essentially that of a particle, like a baseball, being thrown into the air. The most basic equation for kinetic energy is KE = ½ (mass) (velocity)^2. Therefore, the kinetic energy of the forward motion is related to the mass.

Furthermore, for electromagnetic energy the velocity for all EM bursts being "thrown" is the same. Therefore the kinetic energy of the forward motion is related *only* to the mass.

Then, as stated above, the kinetic energy of our EM burst in forward motion is the same as the launch energy. Therefore, the launch energy for any EM burst is primarily related to the mass of the strings to be launched.

For either of these reasons, the Launch Energy required for the EM Burst is dependent on the mass of the energy strings which will be launched.

(New)
**These logical arguments and conclusions are something I came up with on my own. I did use the classic Kinetic Energy equation, yet I applied it to my new concepts of energy strings in a sphere, plus the idea of arranging strings before launch. Therefore much new, yet using classical principles.

106. Therefore both the Inherent Energy and the Launch Energy are related to the mass of the strings. Similarly, the pulsation frequency and the kinetic energy of forward motion are both related to the mass of the energy strings. The processes are slightly different, yet both are initially related to the group of energy strings which will be launched. (New)

107. A specific correlation between the mass of the energy strings and the pulsation frequency is as follows: An EM burst with greater mass will pulsate at a faster frequency. Conversely, an EM burst with less mass will pulsate at a slower frequency. (The exact processes are discussed in a later chapter).
(New)
*This is another insight, based on logic and from previous insights.

108. The mass of the energy strings will NOT change the forward velocity of the EM burst. However, a greater mass of strings in an EM burst will have greater kinetic energy.
(Partially New)
*This is based on classical physics, just applied to new understanding of the photon with energy strings.

109. There is also a correlation between mass and the diameter of the EM burst: The bursts with the greatest mass will have the smallest diameters. Conversely, the bursts with the least mass will have the largest diameters. (The specific cause for diameter is explained in a later chapter).
(New)
*This is really two insights, summarized into one point. First I have discovered the cause of diameter for EM burst, then I have shown that mass is the main cause of diameter, when then leads to this summary statement

110. In total, the mass of the EM burst will determine the following traits.
The EM burst with greater mass will:
a. Have greater energy
b. Have smaller diameter
c. Have greater frequency / will pulsate faster
d. Have greater kinetic energy of forward motion

The EM burst with less mass will:
a. Have less energy
b. Have larger diameter
c. Have lower frequency / will pulsate slower
d. Have less kinetic energy of forward motion

The Process of Launching the EM Burst

111. The actual launch of the EM burst is a two-step process: 1) diverting an initial amount of energy to the energy strings, then 2) adding enough energy to the strings so that they can break free from the electron.
(New)
*This is restated and clarified from above.

112. Before an EM burst can be launched, the energy strings must be given a certain amount of inherent energy. This is essentially building the EM burst prior to launch.
This means that a group of energy strings are arranged in such a way as to have: a certain number of strings, with certain thicknesses, and arranged in a particular way. This arrangement will become the EM burst when actually launched. (New)

113. Just as we can build different size rockets to send into space, we can create different amounts of size EM bursts to be sent into space.

Specifically, we are creating EM bursts with various numbers of strings and thicknesses. Altogether, this creates different groupings of strings, any one of which has the potential of being launched into the air.

(New)

114. However, only a few possible arrangements of energy strings can be created from any one electron or molecule. Thus, although an electron may produce varying EM bursts of different energies and frequencies, only a few arrangements will actually be launched into the air.

(Partially New)

115. Remember that the energy strings in the electron system can combine, separate, and generally rearrange. Therefore, the energy strings can easily shift to another arrangement before being launched.

In practical terms, this means a moving energy target for launch values. Just when we add enough energy to launch a group of energy strings, the group of energy strings rearranges, such that we need a new value for launch energy.

Thus, in order for an EM burst to actually be launched, we have to have the right amount of launch energy for the energy strings as currently arranged, at a particular moment in time. Yet because the energy strings can quickly rearrange prior to getting the proper launch energy, we may not actually launch an EM burst.

(New)

116. Therefore, the basic process of launching an EM burst occurs in the following steps:

   a. Energy must first be diverted into the Inherent Energy. This means that the energy is diverted to the energy strings rather than the motion.

   b. This will define the frequency of pulsation when the burst is launched.

   c. This amount of inherent energy then defines the mass.

   d. The amount of Launch Energy required to emit the burst is then based on that mass.

   e. When enough energy has been diverted to the energy strings to launch the strings of that mass from the electron, then the EM burst will be emitted.
   (New)
   *This is a major breakthrough. Understanding the exact process of emission of electromagnetic energy is a big deal, something that has never been done before.

117. The amount of energy required to emit a burst of electromagnetic energy is known as the "Threshold Energy".

This threshold energy is the total energy of the inherent energy and the launch energy.

This threshold energy is actually a "threshold percentage". This value is the percentage of energy diverted to the energy strings (both the inherent energy and the launch energy) as compared to the total energy of the electron system.
(New)
*This simple set of statements provides a lot of insight into the EM emission processes.

Motions of EM Burst in the Air

118. The direction of forward trajectory of the EM burst is determined primarily by the direction the energy strings are flowing when launch occurs. (New)

Energy Strings versus Electron Motion
(Driver Strings and Non-Driver Strings)

119. Most of the motions of the electron are actually driven by energy strings. (The details are discussed in a later chapter). (New)

120. Energy strings in an electron can have two roles: Driver Strings and Non-Driver Strings. The "driver strings" control the motions of the electrons. The "non-driver strings" extend through the air. (New)

121. The driver energy strings create the various motions of the electrons, including spin, vibration, and forward motion. The direction of the flow of the driver energy strings will determine the direction of each motion. The amount of energy contained in the driver strings will determine the energy and speed associated with each motion of the electron. (The details are discussed in a later chapter). (New)

122. The non-driver energy strings extend outward from the electron into the air. These non-driver energy strings are what we measure as the electrical field and the magnetic field. (New)

123. The driver energy strings and non-driver energy strings can change roles. Thus, driver energy strings can become non-driver energy strings, and non-driver energy strings can become driver energy strings.
   How easy it is for energy strings to change their roles will depend on the actions and status of the electron at the time. In a stationary orbital changing roles is less likely. In electrical current and when absorbing electromagnetic energy the change of roles is much easier. (New)

124. When we talked earlier of the energy of the electron system being in two areas (electron motion and in energy strings) what really exists is driver energy strings (which create the motions) and non-driver energy strings (which creates the fields we can measure). Thus we can restate earlier concepts as follows:

    a. The "transfer" of energy from "electron motion to energy strings" is really the process of energy strings changing roles from driver to non-driver.

    b. The "total energy of the electron system" is in fact a combination of driver energy strings and non-driver energy strings.

    c. The "Energy Percentage" is really the percentage of energy in non-driver energy strings as compared to all the energy strings in the electron system.

(New)

Types of Energies in the Strings and Electron System

125. There are many energies involved in the electron system, and in the launching of EM bursts. Therefore it is good to look at all of these types of energies and see how they interrelate.

126. Electric and Magnetic Fields: What we commonly know as the electric field is actually a set of electric energy strings. What we commonly know as the magnetic field is actually a set of magnetic energy strings.
(New; yet restated from above)

127. Energy Strings: The energy strings are containers of energy. They have both mass and energy. Thus the energy strings with greater mass have greater energy, and vice versa. The energy strings are the simplest form of the Energy = Mass relationship in the universe.

Energy strings remain attached to the electron, until enough energy is diverted to allow them to break free.
(New; yet restated from above)

128. Driver Strings and Non-Driver Strings: An energy strings can either be a driver energy string or a non-driver energy string. The driver energy strings drives the motions of the electrons. The non-driver energy strings extend into the air and will be measured as a field.

(New; yet restated from above)

129. Inherent Energy and Launch Energy: The "inherent energy" is the grouping of energy strings that will become the EM burst. When the EM burst lifts off, this is the group of energy strings which is the essence of the EM burst, with its inherent energy, and resulting pulsation frequency.

The "launch energy" is the energy required to launch the group of energy strings into the air. The launch energy is the extra energy needed so that the energy strings can break free from the hold of the electron.

Remember that the energy strings can rearrange themselves easily. Therefore there may be many possible "inherent energy" groupings. Each grouping will be a different energy photon when launched.

Furthermore, because the energy strings rearrange themselves easily, a particular grouping may rearrange itself before launching energy can be applied.

130. Threshold Energy: The "Threshold Energy" is the amount of energy required to launch a particular size EM burst from the electron.

We can now see that the Total Threshold Energy is a combination of the following:
1. Inherent Energy of Non-Driver Electrical Energy Strings
2. Inherent Energy of Non-Driver Magnetic Energy Strings
3. The Launch Energy required to Launch the particular mass of this grouping of non-driver electrical strings and magnetic strings.

(New; combined new concepts)

A More Sophisticated Understanding of Launching EM Burst

131. Now we can understand the process of launching an EM burst on a much more sophisticated level. An EM Burst will actually be launched when the following process occurs: (New: Very New)

    a. Non-driver energy strings (both magnetic and electric) arrange themselves in a particular way. This becomes the inherent energy.

    b. The electron then gains additional energy strings from an external source.

    c. These energy strings usually take on the role of Driver Energy strings first. Thus, the electron will speed up, vibrate faster, change directions, or change location.

    d. After a while, the energy strings migrate internally. Many of the new driver energy strings become non-driver energy strings.

    e. Thus, the new energy is finally shifted from the motions of the electron to the set of main electron strings, as the new energy strings change roles from driver to non-driver.

    f. Now the main energy strings (non-driver) have enough energy to break free from the electron. The main group of energy strings finally has their "launch energy".

    g. It is at this point that the burst of electromagnetic energy is actually launched.

    h. A photon of a particular energy is emitted, and the electron's motion is observed to be less energetic.

(New: Very New)
*This version of details explains quite a lot related to this topic.

Chapter 8: EM Bursts from Power Lines and Antennas
Summary of Concepts

Primary Concepts

132. All sources of EM emission, including power lines, transmission antennas, and vibrating molecules, are essentially the same physical entities. These entities involve alternating current, alternating electrical fields, and the transfer of energy strings.
    (New)
    *I will show how everything is essentially the same as a power line. This is a new way of looking at things, and it provides a common backdrop for all sources of EM emission.

133. Therefore, the basic process for emission of EM bursts from all sources (including transmission antennas and vibrating molecules) is essentially the same as EM bursts emitted from a power line.
    (New)
    *I will show how everything is essentially the same as a power line. This is a new way of looking at things, and it provides a common backdrop for all sources of EM emission.

134. Electron motion is created by the driver energy strings.
    (New, repeated from earlier)

135. Electrical fields and magnetic fields are actually non-driver energy strings.
    (New, repeated from earlier)

136. Energy strings within the electron system are dynamic. They change roles from driver to non-driver. They combine, disassemble, and rearrange. The particular arrangement of strings at any moment will determine the amount of motion, direction of motion, and strength of the fields. However, because the arrangement of strings is dynamic, the exact motions and fields can change from moment to moment.
    (New, repeated from earlier)

EM Burst Emitted from a Power Line – in Brief

137. Power lines are designed to carry electrical current, and are not intended to emit bursts of electromagnetic energy. However, EM bursts are often emitted from power lines, particularly power lines which carry very high voltage.

138. The basic process for the EM burst to be launched from a power line is as follows:

   a. External energy is applied to the electrons. This energy is applied in the form of magnetic energy strings delivered from a magnet in the generator.

   b. The additional energy strings cause the electron to vibrate faster and to move forward, thus creating electrical current.

   c. The magnet in the generator rotates the opposite direction, which draws the electron to travel in the reverse direction. Thus, an electrical current is created in the opposite direction.

   d. This process repeats, quickly, for as long as there is a power source connected to the generator. Thus, alternating current is created and delivered through the power line.

   e. When the electron moves forward, the energy strings extend upward, and when the electron moves backward the energy strings extend downward. Thus an alternating electrical field is also created along with the alternating electrical current.

   f. These electrical energy strings will generally remain attached to the electron, unless given enough additional energy to break free.

   g. At the same time, there are driver energy strings (which create the motion of the electron) which become non-driver energy strings (which we register as the field).

   h. Also, some of these non-driver magnetic strings can convert into non-driver *electrical* strings.

i. When enough energy strings have been diverted from the motion of the electron (driver strings) to the fields (non-driver strings), and when enough of those non-driver magnetic energy strings convert to non-driver electrical strings, then the entire group of non-driver energy strings will lift off.

Thus, a burst of EM energy is created and launched.

(New)
*Many aspects here are new, a few aspects are known, but even those are not known in terms of energy strings.

Creating Electrical Current: Detailed Understanding

139. The traditional understanding of creating electrical current is as follows:

A magnet exists in a generator. This strong magnet produces a magnetic field, which is strong enough to push the electrons in a nearby wire. The electrons, having been pushed forward, thereby create an electrical current.

An alternating current is created when the magnet is rotated 180 degrees. When the magnet is rotated, this magnet pulls on the electrons in the wire. Thus the electrons flow back to the generator, making a reverse current.

Because the magnet is continuously rotating, the electrical current is created in each direction, repeatedly. Thus, alternating current is produced in the power line.

A more detailed and accurate understanding of electrical current can be explained in the summary points below.

140. A good electrical conductor exists when the electrons are able to move easily from the outer orbital.

141. A strong magnet is set up to rotate near this wire. The magnetic field extends from the magnet, to the nearest electrons on the wire. This provides an external kick to the electrons.
   More specifically, some of the magnetic strings leave the magnet and enter the electron system. Most of these energy strings go in the role of driver strings. Thus, those new magnetic strings actually drive the electron forward. (New)

142. Because these particular electrons are loosely held by the atom, the electron moves straight ahead, like a car being driven forward. This provides a small forward current.

143. However that this electron does not travel all the way. Instead, this electron bumps into the next electron, which sends that electron further down the line. This second electron bumps into the third electron and so forth, through the wire.
   Thus what we know of as an "electrical current" in a power line is actually a sequence of electrons bumping.
   And the "flow" we see of the electrical current is actually the flow of electrons bumping, like the sequence of dominoes falling, and not the actual flow of electrons themselves. (New)
   *This is a new discovery I made, and was first presented in my book Introduction to Electrical Power. The idea of electrons bumping is very different from electrons flowing a long distance.

144. More specifically, the process of electrons bumping is actually the transfer of energy strings from one electron to the other. This is very much like batons being passed person to person in a relay race. (New)
   *This is very new, and it goes even further than electrons bumping.
   This really explains the process.

145. When energy strings are transferred from on electron to the next, these energy strings go first into the role of driver strings. Thus, the second electron has additional energy strings, in the driver role, and is therefore driven forward. (New)

146. Of course, when the energy strings leave the first electron for the second electron, the first electron has fewer energy strings. Therefore the first electron has less overall energy.

147. Energy strings are transferred from the non-driver (field) strings. This is because these are the strings which extend furthest out from the electron. (New)

148. Furthermore, because the energy strings have left the field, the electron will shift more of its energy strings from driver to non-driver. This means that there will be fewer energy strings in the driver role, and therefore the electron will slow down. (New)

149. Thus, what we see as an electrical current is in fact a series of electrons bumping, where each electron pushes the next electron.
    The specific act of "pushing" or "bumping" is due to the transfer of energy strings.
    The first transfer occurs when magnetic strings leave the magnet in the generator, and join the electron system of the first electron. The additional energy strings drive the electron forward, where it "bumps" into the next electron. At this point, energy strings are transferred from the first electron to the second electron. This process occurs from electron to electron, down the lattice of atoms, thereby creating what we observe as an electrical current. (New)
    **All of this is new. It is summary of new items above. But all a very big deal in really understanding electrical current!!

Electrons and Holes Above Atoms

150. Many scientists refer to the outer electrons in a metal as a "sea" of electrons, where all electrons seem to float endlessly around. There is some truth to this, but we can expand our understanding even more.

The system of electrons in a good metal conductor is really a system of electrons and holes. Every metal atom has an electron in its outer orbital. If this electron is knocked out of its orbital, there will be a hole, and the atom will want to have an electron there.

Thus, when electrons bump: the first electron hits the second electron, pushing it out of its spot. This second electron is then part of the electrical current. However, the first electron has now *taken the place* of the kicked out second electron.

Therefore, it is true that electrons in a metal seem to float around and take the place of other electrons in the metal conductor. However, to more specific it is not really a "sea" as it is an electron kicking out another electron, and at the same time replacing that electron in that location.

(New)
*This is also a really new idea

Voltage in Power Lines

151. Voltage is the energy of electrons. When we measure voltage we are in fact measuring the overall energy of trillions of electrons at one location.

(New)

*I first came up with this understanding in my book "Introduction to Electrical Power", and it really is new.

*I am the first person to see what voltage really is: the overall energy of electrons at a location. While it is true that some scientists know that "eV" is the energy of the electron, look in any physics text or website and they will talk of voltage as "potential" or a "water hose" or something equally silly.

Thus, this discovery of mine is a BIG DEAL. Voltage is an important concept. When you understand it accurately, a lot more things make sense.

152. More specifically, voltage is essentially the amount of energy of the energy strings.

For a single electron, the voltage is the total energy of the energy strings in that electron. When we measure voltage of a wire, we are in fact measuring the overall energy of all the energy strings of all the electrons at a particular segment of wire.

Note that I have not decided on this concept: Is the voltage the energy of the driver energy strings only? Or is the voltage the energy of all energy strings (driver and non-driver)? This is a subtlety I have yet to determine.

However, I am more inclined to believe that voltage is just the driver energy strings, which creates the motions of the electrons. (Then the motions of the electrons are what the voltage reading device actually measures).

Generally, the "fields" are measured in other ways than voltage. And for that reason I am inclined to think that the voltage measuring devices are actually measuring the motions of electrons, and not the fields, and therefore voltage is only the driver energy strings. (New)

*The concept of voltage as energy strings of the electron is a new concept, which came to me recently.

153. When energy strings are transferred from electron to electron, only a few energy strings are transferred at a time. Therefore with each subsequent bump, fewer and fewer energy string are making it to the next electron.

This means that each subsequent electron will not move as fast as the previous electron.

Eventually this will result in electrons obtaining only very few energy strings. And therefore the new speed of the electron will be almost nothing. In other words, the flow of electrons, the electrical current, will eventually stop. (New)

*This is very new

154. As a corollary to the above, the previous electron retains some of its additional energy strings. This means that the previous electron has a stronger electric field and magnetic fields than it did in its normal state. This previous electron may also have slightly faster motion than in its normal state. However these will be much smaller and much closer to the normal state after transferring the energy strings.

(New)

155. When we want to send electrical current over long distances, we increase the voltage. This is due to the reasons discussed above.

As stated above, each electron only transfers some of the additional energy strings to the next electron. This means that each subsequent electron will move at a slightly slower speed than the previous electron, eventually resulting in no current being produced at all.

Therefore we use a large voltage at the beginning. Specifically, this means that we apply an enormous amount of energy strings to the first electron. If we begin with an enormous amount of energy strings in our electrons, then there will still be many energy strings which will be transferred to the next electron, even hundreds of miles down the line.

(New)
**This is very new. People have long known that we can apply greater voltages to reduce power loss along a power line, and therefore carry electrical power over longer distances. However, what is really new is explaining the reasons WHY we need greater voltage, and what that means on the microscopic scale.

Alternating Electrical Current

156. The magnet in the generator rotates. Therefore the energy strings of the magnet will flow the other direction. The energy strings of the magnet will then pull on any magnetic strings nearby, including the magnetic strings of the nearby electron. (New)

*Scientists have long known that rotating the magnet will pull on the electrons, creating the reverse electrical current. However, I am now providing the main subtle details, starting with the energy strings of the magnet pulling on the energy strings of the nearby electrons.

157. Because the energy strings of the nearby electron are attached to the electron, the electron is pulled along. This is like pulling a wagon along. Thus, for example, the electron nearest the generator magnet is pulled to the left. (New)

158. Note that the electron closest to the generator also has some of the additional energy strings. This is because not all energy strings were transferred to the neighboring electrons (as discussed above). Furthermore, of all electrons on the power line, the electron nearest the generator has retained most of the added energy strings from earlier.

This means that we have many strings which the magnetic can pull. This is like a stronger set of ropes or a thicker handle attached to our wagon. Therefore, this first electron is pulled along very strongly back toward the generator.
(New)
*This is very, very new.

159. Some of these energy strings will then join with the energy strings of the magnet. In other words, the magnet will regain some of the energy strings which first left it at the beginning of the sequence. (New)

160. Now that one electron has moved, there is a "hole". The nearest atom over would like to have an electron in an outer orbital above that atom. Therefore, the electron one place over will move backward to fill its place. (New)

161. We now have another hole. Thus another electron goes backward to fill that electron's place. This process occurs repeatedly, electron by electron, each one filling the empty space of the other orbital, or being pulled backward directly by energy strings.

This creates a flow of electrons toward the generator. Thus, a reverse electrical current is created. (New)

162. The net result is that all electrons are put back in place exactly as they were before. All electrons are set up in exact position to being the process again.

Therefore, when the rotating magnet comes around again, this magnet can push the nearby electrons, and start the process again. This is like setting up our dominoes again so we can push the first one, and begin another forward flow. (Somewhat new)
*Everyone knows generally about reverse electrical current. However the details, including the idea that everything needs to be put back in place for the process to repeat is a new way of looking at it.

Energy Strings in Magnet Being Replenished

163. The magnet in the generator itself is continually getting replenished.

There will always be new energy strings coming to the magnet in the generator. These energy strings originally come from the power source, such as flowing water or steam. These strings are passed onto the electrons in the turbine, and to the electrons in the axis, and finally to the electrons in the magnet itself.

Therefore, any energy strings which left the magnet, and created the electrical current, will be replaced from energy strings coming from an external power source. (New)

*This is extremely new. This is very new. The whole idea of energy strings being transferred along – from hydropower or coal power to the turbine, to generator, and to the magnet (and of course then to the electrons in the power line) is a VERY BIG NEW idea.

Yet, of course, it explains quite a lot.

164. General Principle of Energy Transfer: The transfer of energy in many situations is actually a transfer of energy strings (primarily magnetic or electric energy strings) from electron to electron, or from atom to atom.

(New)

*This simple statement really is another Big Idea. Within this General Principle of Energy Transfer we can explain a great majority of energy transfer processes.

Pair of Magnets

165. Note that if we have two magnets, rather than a magnet and a power line, the situation is slightly different.

In a pair of magnets, the magnets join when the energy strings join. Then when you pull them apart, the joined energy strings break apart, and each magnet has its own set of energy strings once again. Thus, there is no replacement needed; it is simply a matter of joining and unjoining.

(New)

*This is a new concept of how magnets actually work. Therefore it is a new concept. Also note that most of this will be discussed in a later chapter, we talk of electrons joined as a part of electron constructive interference.

Energy Fields on Power Lines

166. Any power line which carries alternating current will also produce electric and magnetic fields. As stated earlier, these fields are actually non-driver energy strings which extend above the electron.

167. The non-driver energy strings exist without a current. Thus, a small electric field and a small magnetic field will always exist, even without any current.

168. However, we observe a stronger electrical field and stronger magnetic field when electrical current is created. Similarly, with a faster electrical current, we tend to see stronger electrical fields and stronger magnetic fields. This can be explained in terms of additional energy strings.

   Recall that the electron will first absorb additional energy strings. Some of these energy strings will become driver energy strings, and thus propel the electron forward. This creates the electrical current. Other additional strings will become the non-driver strings, which then creates the observable fields.

   Thus, we observe electric and magnetic fields with the creation of electrical current precisely because of the additional energy strings. Some additional energy strings drive the electron forward, while other additional energy strings create the stronger magnetic field. (New)

169. A stronger energy field is associated with a faster moving electron. Again this can be explained by the additional energy strings. When we add a large amount of energy strings to the electron system, then we will have many additional driver energy strings, which thus makes a faster electrical current). This also makes many additional non-driver energy strings, which makes a stronger energy field. (New)
   *New, because of explanation

170. Magnetic fields are similar to electric fields in that:
   a. They are actually a set of non-driver energy strings
   b. They exist within the electron system regardless of current.
   c. When additional magnetic strings are added to the electron system, the electron will move faster, and the magnetic field will be stronger.

171. The magnetic field is also perpendicular to the electrical current. However, the field is also perpendicular to the electrical field as well. (This has been well established).

Emission Process from Power Lines, in Detail

172. Although power lines are designed to carry electrical current, these power lines can emit bursts of electromagnetic energy as well. This is particularly true of power lines with very high voltage.

173. Most EM emission processes are simply variations from EM emissions from the power line. Therefore, by first understanding the process for EM burst from a power line, we can then learn the variations of EM emission from all other sources. (Partially New)
   *I saw the correlations myself, after studying multiple sources of emission. If anyone else had seen this before I did not know of it. Furthermore, some people I have talked to are not aware of how all the emission processes are really versions of the power line. Therefore my understanding and seeing the commonalities is new to many scientists.

174. External energy is applied to the electrons. This is done by energy strings from outside the electron entering the electron system and being absorbed by the electron system. (Partially New)

175. Some of the additional energy strings become driver energy strings. This causes the electron to vibrate faster and move forward faster.
   The other additional energy strings will be non-driver energy strings, which then make the measured energy fields much stronger.
   (New, mostly repeated from earlier)

176. The non-driver energy strings (the energy fields) are the actual energy strings which will eventually launch from the electron and become independent bursts of electromagnetic energy.
(New, summarized from earlier)

177. Most of the time, these non-driver energy strings will remain attached to the electron. It is only when the non-driver strings (energy fields) have enough energy to break free from the pull of the electron will the energy strings lift off from the electron and become an emitted EM burst.
(New, summarized from earlier)

178. The energy strings within the electron system exist in a highly dynamic situation. Driver string become non-driver strings, and vice versa. Energy strings join, break apart, and rearrange.
   Therefore it is possible at any time for enough energy from the driver strings (motion of the electron) to be diverted to the non-driver strings (the fields) in order to create and launch a burst of electromagnetic energy.
(New, summarized from earlier)

179. In addition, magnetic energy strings can convert to electric energy strings. This process diverts additional energy from the magnetic field to the electric field. This is important for our inherent energy and our launch energy. (New)

180. The actual launch of the EM burst from a power line will occur when the energy strings of the both energy fields have enough energy to actually break free from the electron system.

We can state this in several ways:

a. Enough energy has been diverted from the motion of the electron to the energy fields, such that the energy fields have enough energy to break free from the electron.

b. Enough driver energy strings have become non-driver energy strings, in order to provide enough energy to the existing non-driver energy strings to break free.

c. Starting with a group of non-driver energy strings, with a particular total mass and total energy (the inherent energy), we then add enough additional energy (launch energy) by diverting strings from the motion of the electron to the fields, such that this original grouping of non-driver energy strings can be launched.

(New, Restated)
*This is an important summary, of how the EM burst is actually launched.

181. A specific burst of EM energy (specific inherent energy and pulsation frequency) will be emitted only when the Threshold Percentage has been reached. This means that the amount of energy in the fields, relative to the total energy of the electron system, allows a particular grouping of energy strings to break free and become an independent entity.

Stated another way: a specific inherent energy and pulsation frequency of EM burst will be launched only when the non-driver energy strings have enough energy relative to the total energy within the electron system to launch a particular set of non-driver energy strings.

This is the specific Threshold Percentage, which will launch an EM burst of a particular inherent energy and particular pulsation frequency.

(New, Restated)
*This is important to remember.

Power Lines, EM Bursts, and Power Loss

182. The process above explains how EM bursts can be emitted from a power line. However, the purpose of a power line is to carry electrical current, not to emit electromagnetic energy. Therefore this is actually a type of power loss.

183. It has long been known that when EM bursts are emitted from a power line this results in power loss. Now we understand why.
   When an EM burst is launched, there are energy strings which have left the electron system. Therefore, the electron has fewer energy strings, and thus less overall energy.
   Consequently, we have two results: EM bursts are emitted, and the remaining electrons have less energy. Thus we have power loss along the wire. (New)
   *This is a new understanding of what happens. Just simple, clear.

184. EM bursts are more likely to be emitted from power lines with greater voltage. This is due to two concepts: a) the nature of voltage, and b) the threshold percentage.
   Remember that voltage is in fact the amount of energy strings in the role of driver strings in the electron system. Therefore, when we increase the voltage, what we are doing is adding more and more energy strings to each electron.
   Then remember that in order for any group of non-driver strings (fields) to launch as an EM burst, that group of strings must have enough energy to break free.
   Thus, if we add a great amount of voltage we are in fact adding a large number of energy strings. These strings divide up into two roles: driver or non-driver energy strings. If we add a huge amount of energy strings to the system, then many of those energy strings will join the existing non-driver (field) strings. The net result is that the non-driver energy strings may easily have enough energy to break free.
   Therefore, by adding large amounts of voltage we may easily be creating more EM bursts from the power line.
   (New)
   *This is a new understanding of what happens. Just simple, clear.

185. There are ways to minimize EM bursts emissions from power lines, including barriers of electrical insulation. These out materials will deflect many frequencies of EM bursts. Like deflecting a soccer ball, the material essentially puts the electromagnetic energy back into play, thereby minimizing loss of energy strings.

Radio Transmission Antenna

186. The radio transmission antenna is essentially the same as our power line. In both we have a good conductor, which carries alternating electrical current. In both the emission of EM bursts can be created.

187. The main difference between the power line and the transmission antenna is the purpose: in the power line we do not want to emit EM bursts because this is a power loss. Yet in the radio transmission antenna we do want to emit EM bursts because we want to send signals through the air.

188. The other differences are technical. These include:
    a. Power lines are lined or laid horizontally, whereas most transmission antennas are installed vertically.

    b. Power lines use a set frequency alternating current of 60 Hz. Transmission antennas use much faster frequencies of electrical current, and the range of possible frequencies varies with application.

    c. Power lines use high voltages to carry electrical power long distances because of the gradual power loss along the way. Transmission antennas use high voltages to emit more EM bursts per second, thereby sending a stronger signal.

*These concepts are well known. Nothing new here. I am just stating for clarification.

Frequency of EM Burst Emitted from Transmission Antennas

189. When we desire to emit an EM burst with a specific frequency of pulsation (inherent energy), there are two main factors to consider: frequency of the electrical current, and height of the transmission antenna.

190. If you create an alternating electrical current which has the exact frequency of the EM burst you wish to emit, then you are more likely to emit an EM burst of that frequency.

Similarly, if you create alternating current which as frequencies which are exact multiples of the frequency you wish to emit, then you may emit the desired frequencies.

191. If the height of your transmission antenna is exactly the same size as the wavelength of the EM pulsation cycle you wish to emit, then you are more likely to emit an EM burst with that wavelength. This of course means that the EM burst will have the corresponding frequency and inherent energy.

192. You can also build an antenna which is exactly ½ or ¼ the size of the wavelength of EM you wish to emit. The exact size of ½ and ¼ works precisely because of the detailed cause of the wave pattern. As long as we create the maximum amplitude, which will become the crest of our wave, at the right distance on our wire, then when the EM burst is launched, and then the EM Burst will take care of itself.

In other words, we do not have to go through a complete cycle, as long as we reach the maximum amplitude at the right length along the wire.

(Partially New)

*The engineers observed that you can do the fraction antennas of ½ and ¼. However, they did not know why. I have figured out exactly WHY this works, AND it is related to the cause of the wave pattern (which is also one of my discoveries).

EM Bursts are Independent After Being Emitted

192. After the burst of Electromagnetic energy has been emitted, the burst is completely independent, and it has a life of its own.

193. The mechanism of the EM Burst is completely different from the mechanism which emits the energy. Once in the air, the EM burst operates in a special way. (The details are described and illustrated in later chapters).
    (New. This is a clarification to understand better)

Creation and Perpetual Cycles of the EM Bursts

194. A concise summary of the creation of EM bursts, followed by the perpetual cycle of EM bursts as independent entities, is as follows:

    a. Electrical and Magnetic energy fields start as being attached to an electron.

    b. Additional energy strings are added to the existing energy strings. Some additional energy strings go into the motions of the electron, while the other energy strings become attached to the field energy strings.

    c. When enough energy has been diverted to the energy fields (the Threshold Percentage) then the energy fields will launch off the wire.

    d. Once the energy is launched from the wire, the EM Burst is independent, with a life of its own.

    e. The EM burst will pulsate based on its own internal mechanism, essentially forever.

    f. The frequency of the EM burst is determined primarily by the frequency of the inherent energy. This is the energy of the field (the non-driver energy strings) in the electron when on the wire.

g. Other factors which affect the frequency of the EM Burst include the frequency of the electrical current, and the height (or length) of the wire.

h. The direction of forward travel of the EM burst is determined by the direction of the energy field at the time of launch.

i. As long as the fields have been launched, an EM burst is created. It does not matter where in the cycle of alternating current the electron is, if the fields have been launched off the wire, the launched fields will become a burst of electromagnetic energy.

High Voltage and Emission of EM Bursts

195. Increasing the voltage of an electron will increase the motions of the electron. This is because increasing the voltage is in fact adding energy strings to the electron system, and because most of those strings go into driving the motions of the electron. (Partially New)
   *Basic concept with voltage is known, but what is new is understanding this in terms of energy strings.

196. Increasing the voltage of the electron will also increase the energy fields. This is for two reasons: first, some of the additional energy strings will join the non-driver strings, thus increasing the strength of the field. Second, the non-driver strings divert to driver strings over time, which thus increases the energy field. (Partially New)
   Increasing the voltage of an electron is also more likely to cause that electron to emit a burst of electromagnetic energy. This is because the non-driver energy strings have more energy, often enough energy to break free from the electron system. (Partially New)
   *Basic concept with voltage is known, but what is new is understanding this in terms of energy strings.

197. Increasing the voltage of an electron allows the electron to emit higher energy EM bursts. This is because the amount of energy in the EM burst depends on the energy in the grouping of strings. Thus, adding more strings to the electron system allows the grouping of non-driver strings to have more energy, and when emitted this becomes a higher energy EM burst. (Partially New)
   *Basic concept with voltage is known, but what is new is understanding this in terms of energy strings.

198. Increasing the voltage of an electron allows the electron to assemble different groupings of energy strings, which will then allow for different inherent energies (and different pulsation frequencies) of EM bursts to be emitted.
   (New)
   *This is a new concept, explaining exactly how increasing the voltage can cause increase in speed followed by possible EM burst.

199. Increasing the voltage of a wire generally means that millions of electrons will have greater energy. This means that more electrons can emit EM bursts at the same time. The net result is what we know of as a greater intensity or stronger signal. (New)
   *Understanding what power and intensity is for EM emissions was not really understood until I explained it (above, previous chapter). Plus, I now put energy strings as voltage, and coming from magnet across the diameter of the wire. Thus, altogether there is much in this statement that is new.

Chapter 9: Creating EM Bursts from Molecular Vibrations
Summary Points

Overview

200. The essence of a molecular bond is that the electrons are shared among two atoms. This connects the atoms together.

201. A molecular bond is traditionally viewed as electrons being shared by atoms. A more detailed traditional view states that a chemical bond is formed when atomic orbitals overlap. I supplement these views with other models, culminating in the model I call the "Molecular Orbital System".

202. Molecular vibration is the process of the two atoms in a molecular bond expanding and contracting, repeatedly.

203. The basic process of EM burst emission from a molecular vibration is as follows. Note that his is the traditional view, prior to the discoveries in this book. This basic process will be supplemented with additional details for more accurate understanding.

   a. The molecular bond in its normal state has electrons with low level of energy. This energy drives the vibration of the molecular bond, and drives it at a certain frequency.

   b. Additional energy is added to the electron. This is usually from the absorption of high energy electromagnetic energy photons. This can also be from impact from subatomic particles.

   c. When this energy is absorbed by the electron, the electron moves at a higher energy. Because the electron is part of a molecular bond, this higher energy is seen as an increase in molecular vibration.

   d. The electron prefers to be in its normal energy state. Therefore, the electron gets rid of its excess energy. This energy is most commonly released in the form of a burst of electromagnetic energy.

   e. This is a "subsequent" burst, which is less energy than the amount of energy in the photon which was absorbed.

f. Thus, the electron gets rid of most of its excess energy, through the burst of electromagnetic energy. An independent entity of electromagnetic energy is sent through the air.

Conversely, the electron has less energy, and therefore will slow down. This will also cause the molecular bond to vibrate at a slower pace.

204. There are several new models which explain molecular bonds and molecular vibration. These models are:
   a. Molecular Bond as Power Line
   b. Molecular Bond as Pair of Parallel Wires
   c. Molecular Bond as Molecular Orbital or Looping Racetrack
   d. Molecular Bond as a Set of Molecular Orbitals.

(New)
*All of these models are new. Although there are some similarities to previous scientific models, my models have some differences. particularly in the underlying physical reality behind the geometries.

Molecular Bond and Vibration as Power Line

205. Any molecular bond is essentially the same as a power line. Therefore much of what we have learned regarding power lines can apply to discussions of molecular bonds.
   (New)
   *This is a TOTALLY NEW IDEA. The idea that a molecular bond is essentially a power line is a new concept, and will help much in our understanding.

206. In the simplest new model of the molecular orbital, the pair of electrons exist on an imaginary line.
   The "bond" exists because both electrons are on the same line, and this line connects both atoms.

207. The vibration of the molecular bond in this model exists because the electrons move back and forth, always in opposite directions.
   As the electrons move toward the atoms, the atoms move further apart. This is expansion. Then as the electrons move toward each other, the atoms move in closer. This is contraction.
   This process continues forever. Thus the vibrational motion is created.

208. Alternating electrical current in the orbital is also created at the same time. By definition, when an electron moves to the right, an electrical current is created to the right. When an electron moves to the left, an electrical current is created to the left.

Therefore, each election is actually producing alternating current as it travels toward the atom and then back toward the center. This is very similar to the electron on the power line moving forward and reverse.

(Partially New)

*This is new in by pointing out that the electron in the orbital is creating electrical current when moving (as in a wire), and second by pointing out that the electron in the orbital creates alternating current.

209. Furthermore, there are usually two electrons in molecular bond. Therefore, we have two independent creations of alternating current in every molecular bond. Note that this principle is true for any of the new models of molecular bonds. (New)

*This is also new. And can be observed in all models I have presented.

210. Energy fields are also observed in the molecular bond.

As discussed earlier, electrons have driver strings (which cause the electron motion) and non-driver strings (which are measurable as energy fields). Therefore, each electron in the molecular orbital will exhibit an energy field. This field will also become stronger as more energy strings are added to the electron system. (This will occur when the electron absorbs energy). (Partially New)

211. Alternating energy fields are also observed with the vibrating molecule.

In this model each electron moves right, and left, and right again, thereby creating a small alternating current. We also from previous discussions that the energy field changes direction as the electron changes direction. Therefore, as the electrical current moves right, the electrical energy field flows upward, and as the electrical current moves left, the electrical energy field flows downward.

This is how alternating electrical fields are observed, for each electron, the molecular orbital.

Of course this field will become stronger and easier to observe as more energy strings are added to the electron system.

Molecular Bond and Vibration as Parallel Wires

212. A more accurate depiction of molecular bonds would be two parallel wires. In this model we have two parallel lines, and each electron is on a separate line. This much the same as if we had two separate wires connecting the atoms rather than just one wire. (New)

213. There are noticeable additions to this model. First, we have two wires rather than just one. Second, each electron travels all the way to both atoms, rather than just to one atom. Notice that the electrons do not interact, nor do they arrive at the atom at the same location. (New)

214. As in the earlier model, the electrons travel in opposite directions. However this time they are on parallel wires as well. Thus, for example, electron 1 travels from left to right on its wire, while electron 2 travels from right to left in its wire. (New)

215. Vibration in this model is created essentially the same way as in the previous model, yet becomes much more pronounced.

As stated earlier, vibration in the molecular bond is a series of expansion and contraction of the two atoms. In this model of the molecular bond (as in the previous one) electrons are traveling in opposite directions.

Thus, as electron 1 enters atom A, while electron 2 enters atom B, the atoms are pushed apart, and the molecular bond expands. Then, as the electrons head in the opposite directions (both electrons leave the atoms and head for the space between them) the molecule will contract.

Then of course electron 1 enters atom B, while electron 2 enters atom A. This again causes an expansion, though pushing of each atom is done by different electrons. The electrons then head back to the center, which causes a contraction.

This is the basic process alternating paths of electrons, each electron on a different wire, which causes the molecule to vibrate repeatedly.

Of course, when additional energy is added to this system, each electron moves faster, and thus the frequency of this expansion and contraction (molecular vibration) is much faster. (New)

*This is very new. The entire set of pictures really shows how this works. It explains a lot. And it is all my model.

217. As with the previous model, each electron in this model creates its own alternating electrical current. That is, each electron moves left to right, left again, and right again. Thus, each electron, and each bond, carries a small alternating current.

Of course there are two separate bonds, with two separate electrons, creating two alternating currents.

(Partially New)
*New, but similar to New insight above.

218. Also as with the previous model, each electron in this model creates its own alternating electrical field.

The process is the same as described above: non-driver energy strings create the electrical field. The direction of the electrical field is correlated with the direction of the electrical current.

And when we add energy to the system, we have a stronger electric field (along with the greater motion of the electron). This is more observable, and will be significant when launching EM bursts from the molecular bond.

Also, because there are two independent wires of electrons, we have two independent alternating electrical fields in our molecular bond.

(Partially New)
*New, and valuable, yet similar to New insights above.

Molecular Bond and Vibration as Molecular Orbital

219. A significant advancement in the model of the molecular bond is the is that of the Molecular Orbital.

The "molecular orbital" is a continuous elliptical loop, similar to a race track. This track goes through each atom, and the spaces in between.

Thus: an electron travels across the space to atom A, around the nucleus, and to the other side. The electron continues along the straight portion of the path, to atom B. The electron goes around the nucleus in atom B, to the other side, and along another straight path, where it reaches atom A again, and the process continues.

On this race track are two electrons. Each electron is positioned equal distance from the other. We then see these two electrons travel round and round the track, through each atom, and along the straight paths in between, in a never ending loop.

In total, this is the basic model of the Molecular Orbital.

(New)

*Totally New. A lot of things make sense with this model.

**Note that although the concept of the "molecular orbital" has been used for decades, my model has significant differences. Primarily, my molecular orbital has electrons as particles traveling around the atoms, rather than vibrating energy strings merging together.

My model also allows the electron to have definite position and velocity, rather than a vague range. This is much more aligned with physical reality.

220. The most significant factor of this model is that the electrons actually travel a loop, which includes traveling through each atom.

This is vastly different from electrons which travel back and forth on a single line, as in the previous models.

This also differs from the models of "atomic orbitals overlapping". This is common view of molecular bond by quantum scientists, yet it implies that each electron is essentially in the region of its atom, and not traveling to the other atom.

In contrast, this view of the molecular orbital shows that each electron travels through each atom. Thus, an electron is physically shared by both atoms, as the electron spends some time in both.

This also creates the "bond", an entity which holds the two atoms close together.

(Partially New)

221. You will notice that the molecular orbital, being essentially an ellipse, has two parallel paths.

These parallel portions of the molecular orbital are significant because they help demonstrate electrons apparently traveling in opposite directions. This then helps explain a) alternating current and b) the process of contraction.

222. Another major factor is that the electrons are placed such that they are always traveling in opposite directions at any time.

Just as two cars on a race track can appear to be driving in opposite directions (simply by their relative locations and spacing), the two electrons on our molecular orbital can appear to be moving in opposite directions at any given moment.

It is because these electrons are moving in opposite directions that the expansion and contraction (the molecular vibration) will be created.

(New)
*This is a totally new understanding of how the electrons create molecular vibration!

223. The molecular vibration in this model is created because of the two factors:

1. It is a continuous loop which connects both atoms, and
2. The electrons are spaced appropriately such that they always appear to be traveling in opposite directions.

In brief: when the electrons in the molecular orbital are moving away from each other, the molecular bond expands. When the electrons in the molecular orbital move toward each other, the molecular bond contracts. This expansion and contraction is what creates the molecular vibration.

More specifically, as each electron gets close to their respective atoms, the electrons push the atoms further apart. This creates an expansion. Then as each electron moves along the straight paths in the space between the atoms, the electrons actually get closer together (and the atoms as well). Thus, the molecular bond contracts. This creates the expansion and contraction, and hence the vibration.

(New)
*This is another reason why my model works so well! And it is absolutely my idea.

224. Each electron in the molecular orbital creates a type of alternating current.

Remember that the molecular orbital is essentially an ellipse. When we watch the electron on this ellipse we first see that it travels from left to right. Then the electron goes around the nucleus, and proceeds on the parallel straight path from right to left.

If we just look at those two parallel sections of the ellipse, we can therefore see that the electron first travels one direction, and then in the opposite direction. This is the very essence of alternating current.

(New)

*This is another great feature of the molecular orbital! I am the first to point out how alternating current exists on this ellipse.

225. Each electron in the molecular orbital will also have an alternating electrical field.

As discussed previously, every electron has non-driver strings which are the observed energy fields. As the electron travels around the molecular orbital those energy fields will expand, contract, and expand in the opposite direction.

Furthermore, the exact size of this field at any moment is related to the speed of the electron, as the driver and non-driver strings change roles.

Of course, when energy is added to the system, some of that energy will go into the motion of the electron, while other energy will become part of the field. Thus, the motion of the electron and the strength of the energy field is likely to increase.

These fields are important, because they will become the grouping of energy strings when launched. These are the energy strings which will in fact become the EM bursts emitted from a molecular vibration.

(Partially New)

*Much of this is new to the world, yet it is a combination of points already repeated elsewhere. However, it is also very important to have this here, because the fields created from the electrons on the orbital will become the EM bursts emitted from a molecular vibration.

226. There are two electrons on this molecular orbital. Each electron operates somewhat independently. Therefore, there are two alternating electrical currents, and two alternating electrical fields.
(Partially New)

227. Note that the molecular orbital can be more complex shaped than the simple elliptical orbital. Yet all of the information discussed above and all the principals involved would be the same. Some details of more complex orbitals are discussed in a later chapter.

Molecular Bond and Vibration as Set of Two Molecular Orbitals

228. The most accurate model for the molecular bond at this point in time is the Set of Two Molecular Orbitals.

The model of the "Set of Molecular Orbitals" is similar to the model above of the single molecular orbital. However, in this model each electron travels on its own path, along its own molecular orbital. (New)

*This is very new. It is all the new ideas of my version of the Molecular Orbital, plus the idea of having the two electrons on two molecular orbitals.

229. The details of the Set of Molecular Orbitals model for molecular bond is as follows:

a. There are two molecular orbitals. These are continuous looping paths, from one atom to the other.

b. The specific shape of the molecular orbital can be almost any complex shape, as long as it is a closed loop, and the electron travels through both atoms.

For simplicity in our discussions we will stay with the most basic shape of the molecular orbital: the ellipse.

c. The path of each orbit is to go around the nucleus of the first atom, through the space between the atoms, around the nucleus of the second atom, and continue back to the first atom.

d. Each electron travels on its own orbital. Thus, each electron passes through both atoms, yet the electrons do this on different paths.

e. Each electron travels in the opposite direction on its own path. Thus, while electron 1 is traveling on its path from left to right, electron 2 is traveling on its path from right to left.

(New)
*This is very new. It is all the new ideas of the Molecular Orbital, plus the idea of having two electrons on two molecular orbitals.

230. A molecular bond is created by each molecular orbital. Because an electron travels through both atoms in a continuous manner, and because both atoms share that electron part of the time, a bond is created.

Then because there are two molecular orbitals the bond is stronger. We are connecting the atoms in two separate molecular orbitals. The atoms are sharing two different electrons, and the electrons travel through the atoms in different regions. Therefore the two atoms are much more strongly connected. (New)

*This is very new.

231. The molecular vibration is created because the two electrons are always traveling in opposite directions.

Each electron is traveling on its own molecular orbital, yet they are traveling in opposite directions. The net result is both electrons push the atoms apart, which expands the molecular bond, followed by both electrons pulling the atoms together, which is a contraction. Thus, a molecular vibration is created. (New)

*It is an extension of earlier new concepts, yet it is all new, from me.

232. The specific steps for molecular vibration as caused by a set of molecular orbitals are as follows:

Step 1: When electron 1 (on its molecular orbital path) is in atom A, we see electron 2 (on its molecular orbital path) is in atom B. Thus, each electron is pushing their respective atoms further apart. This creates expansion.

Step 2: As each electron goes around the nucleus, the electron leaves its atom and travels a straight path, heading toward the next atom. Thus, electron 1 starts heading toward atom B, and electron 2 starts heading toward atom A. This creates a contraction.

Step 3: When the electrons reach the other atoms we see that electron 1 is now in atom B, and electron 2 is now in atom A. Thus, again each electron is pushing their respective atoms further apart. Yet the each electron is doing this for the other atom that it was previously. This also creates expansion.

Step 4: As with step 2 the electrons go around the nucleus, and start heading back toward the other direction. Because both electrons do this, in opposite direction, a contraction is created.

This process repeats, causing a continuous vibration.

(New)

*It is an extension of earlier new concepts, yet it is all new.

233. Each electron also creates a type of alternating current is the electron travels around the molecular orbital.

Because the electron is in motion, this is the essence of an electrical current. Then because the electron travels from left to right (on one side of the ellipse) and then from right to left (on the other side of the ellipse) we essentially have alternating current. (New)

*It is an extension of earlier new concepts, yet we are now combing the idea of two parallel wires with a single loop. All of these ideas are new (as stated above), and this new combination is new to the world as well.

234. Each electron also creates alternating energy fields as the electron travels around the molecular orbital.

As stated earlier, the energy fields are actually non-driver energy strings. Both electrical energy strings and magnetic energy strings flow perpendicular to the direction of electrical current, yet also perpendicular to each other.

Also as stated earlier, the flow of the field energy strings is related to the direction of the electrical current. For example, as the electron travels left to right, the electrical energy strings flow upward, and as the electron travels right to left the electrical energy strings flow downward.

These alternating fields will become the group of energy strings in the EM burst when launched.

(Partially New)

*It is an extension of earlier new concepts. Important to restate, just for this model of molecular orbital.

Process of EM Burst from Molecular Vibrations

235. In the stable state of the electron in its molecular orbital, the electron has a certain amount of energy. This energy is observed as a particular amount of motion (which creates a certain frequency of vibration), and a particular strength of the energy fields.

236. If enough energy is diverted to the energy strings of the energy field, then those energy strings can break from the electron, and become an independent EM burst. However, this usually does not happen. Usually there is not enough energy in the electron's stable state for this to occur. Therefore the energy strings remain attached to the electron.
   (Partially new)
   *Extension of previous ideas.

237. The electron can absorb external energy. This external energy usually arrives as an incoming photon.

238. The basic mechanism of absorption is that the energy strings of the photon will join the energy strings of the electron. Thus the photon "disappears" while the electron has obtained all of that energy. Further details are discussed in a later chapter. (New)
   *This is very much new. This is understanding the exact process of absorption. Yet more details are presented in a later chapter, this is just the brief version.

239. Some of the additional energy strings absorbed will become driver strings. This will cause the electron to travel faster, and therefore the vibration process to occur at a much faster frequency. (New)
   *Extension of ideas, and building on concepts.

240. Some of the additional energy strings absorbed will become non-driver (field) strings. In addition, some of the driver strings will diver to non-driver strings. Therefore, for both these reasons, the non-driver energy strings (the fields) of the electron after absorbing energy will be much stronger. (New)
   *Extension of ideas, and building on concepts.

241. A burst of electromagnetic energy will be emitted from a molecular bond when the energy field strings have enough energy to break free from the electron. This will happen after:

    a. Enough additional energy has been absorbed.

    b. A particular arrangement (inherent energy) of energy strings has been grouped together.

    c. Enough energy has been diverted from the motion of the electron (the driver strings) to the energy fields (the non-driver strings) in order for those energy strings to break free.

(Partially New)
*This basic process is a new idea in general. It is a big idea, and worth repeating. Yet it has been stated above.

242. It is possible for a molecular bond to emit one of several frequencies of electromagnetic energy. The specific frequency emitted (the specific energy emitted) depends on:
    a. The amount of energy initially absorbed
    b. The first Threshold Percentage reached
(New)
*This is described in greater detail later.

243. If multiple EM bursts are emitted in sequence, the specific EM bursts emitted depend on:
    a. The total energy in the electron system at any given moment.
    b. The next Threshold Percentage to be reached.
(New)
*This is described in greater detail later.

244. As a consequence of the EM burst emission, the molecular bond vibrates at a slower frequency.

This is easily understood: the electron has fewer energy strings overall because some energy strings have left as the EM burst. Thus, there are fewer driver energy strings to drive the electron forward. Consequently, the electron will travel slower, and the expansion and contraction process will be slower.

(Partially New)
*Slower frequency of vibration after emission is known, and because of loss energy is known. I have put in terms of energy strings, which is new.

Number of EM Bursts Related to Molecular Orbitals

245. Although there are two electrons and two molecular orbitals in the system, only one electron is necessary to emit an EM burst.

246. Because there are two electrons and two molecular orbitals it is possible for the molecular bond to emit two EM bursts at the same time.
    (New)
    *This is extension of earlier concepts and the orbital model.

247. Also, because there are two orbitals each orbital can operate independently. This means that each electron will absorb energy independently, each molecular bond will vibrate independently, and each electron will emit EM energies independently. (New)
    *This is extension of earlier concepts and of the orbital model.

248. Because there are two electrons in two independent molecular orbitals, and because either electron can emit one of several possible frequencies of EM bursts, it is possible for the two electrons to emit two different frequencies of EM bursts at the same time. (New)
    *This is extension of earlier concepts and of the orbital model.

Incoming Photon and Molecular Orbitals

249. It is also possible for one photon to be absorbed by both electrons in the molecular orbital system. This will occur when the photon is approximately the same size as the molecular bond.

    In this situation, the energy strings of the photon will be spread across the molecular bond, and reach both of the electrons simultaneously. Thus, both electrons can absorb the energy strings of the same photon.
    (New)

Additional Factors in the Creation of EM Burst

250. In addition to the basic mechanism for creating an EM bursts from a vibrating molecule, there are a few related concepts:

   a. Frequency of EM burst is related to speed of electron.

   b. Loss of energy in an orbital system is diverted to energy strings and EM burst.

   c. Amount of time a molecule exists in a higher energy state before emitting EM burst is variable, depending on the threshold percentage.

   d. When an orbital system acquires energy, this system can exhibit higher energy through change in speed, shape, and size.

   e. Most molecular bonds can emit one of several frequencies of EM burst; which one emitted depends on the relative percentage of energy in energy fields versus energy in the total orbital system.

251. The rate at which the electron travels around the molecular orbit is a cyclical frequency, and therefore is the major factor in the frequency of the EM burst which is created from that orbital. (Partially new)
   *Scientists from 1900 to 1940 believed that the frequency of EM burst was related to the speed of the electron. However, they discarded this idea for various reasons. I am reviving this idea.
   When you have the complete understanding of energy strings and how EM bursts are created, you will understand that the speed of the electron around the orbital is part of its internal energy (the other part being the energy field), and the total energy strings is divided in some percentage between driver strings and field strings. Therefore the speed of the electron does in fact influence the frequency of the EM burst emitted.

252. The exact amount of energy emitted as an EM burst is also the amount of energy lost to the electron system. These interrelated concepts explain a) how the EM burst obtains its amount of energy from the orbital system, b) how a molecular bond loses energy when an EM burst is emitted, which therefore c) results in the orbital system existing at a lower energy.

253. There is no fixed time for any electron to be in its higher energy state. The electron can remain in its higher state for a short time or a long time.

Also, the amount of time in which the electron is in the higher state is completely independent of the energy of the system.

It is only when the energy of the fields combine in such a way as to reach the Threshold Percentage will an EM burst be emitted, and the electron drop to its lower energy state.

(Partially new, mostly restated and clarified concepts)

254. When the electron in an orbital system acquires energy, this system can exhibit higher energy through change in speed, change in shape, and/or change in size.

255. Increase in speed: When the electron acquires energy, it is really acquiring more energy strings. Some of these energy strings will become driver strings, which will therefore cause the electron to travel faster through its orbital path.

256. Change in Shape of Orbital: When an electron acquires energy strings, and many of those become driver strings, the orbital can change shape.

This is because the energy strings may move the electron in a variety of different directions (depending on the flow of the driver strings), and with different amounts (depending on the number of driver strings in any one direction).

Therefore, the electron becomes pulled in a variety of different directions than before. The net result is a different orbital path. This is how electrons of different energies will create different orbitals.

(New)

*Bohr and Sommerfeld were the men who really figured out that orbitals are different shapes, and that each orbital is due to the specific energy of the electron.

However, I am the first to show how more energy strings, particularly additional driver strings, in their various motions and directions, will create the new path of the electron.

Schematic of EM Absorption and Subsequent Emission

257. In the traditional schematic of EM absorption and subsequent emission, the energy levels are drawn as a series of lines, similar to a series of book shelves.

258. Each level of the schematic usually refers to the possible energies which can either be absorbed by the electron or emitted by the electron.
    a. The higher "shelf" represents a higher energy level.
    b. The "shelf" or "level" system exists because the electron will tend to absorb only certain frequencies, and emit only certain frequencies.

259. The traditional schematic of EM absorption and subsequent emission can be used to represent a number of different situations.
    Note that the author must clarify exactly what the levels are referring to in his diagram:

    c. The author should specific which electron the schematic refers to, because every electron in an atom will have a different set of energy levels in the schematic.

    d. The author should also specific whether the schematic refers to the absorption levels, the emission levels, or both.

260. A single schematic is commonly used to show how electrons can absorb certain energies, then emit certain energies. However, in reality there are two sets of schematics: 1) the absorption schematic, and 2) the emission schematic.
    Two schematics really exist because electrons tend to absorb only certain energies, and tend to emit certain energies. These sets of energies are not necessarily the same.
    Yet for simplicity in understanding the processes and making comparisons, it is useful to merge these two schematics together. This means all levels are shown, for absorption and for emission, though the "emission" lines are not likely to be "absorption levels", and vice versa.

261. The lowest level usually represents the energy level where the electron is stable. The electron does not emit any EM burst on its own.
    Note that the energy level of the lowest level will be different for each electron in an atom.

262. The higher levels represent higher energy levels. In an absorption schematic the higher levels represent the levels to which an electron will absorb energy.

263. Notice that the highest level in any schematic is usually the highest level of energy for that electron in that orbital path *for which the electron will be unstable*. Any additional energy will send the electron to the next orbital. If the electron gains enough energy, it will reach that next level, where it will reside in the next stable orbital.

For example, adding enough energy to send the electron's energy to level 4 will be the highest unstable energy state for that orbit. Yet, if we added enough energy to go beyond that, we have in fact moved the electron to a new orbital, and the electron is actually stable once again.
(New)
*This is my addition to traditional understanding. There is really nothing "new" in the models, but I have clarified the understanding of what we are talking about and representing.

264. Similarly, the emission schematic shows us the energy of the electron temporarily existing at one energy level, emitting and EM burst, then dropping down to a lower energy level.

Energy Level Schematic with Energy Strings

265. With our understanding of energy strings and the process of launching an EM burst, we can now correlate the schematic of energy levels with the actual processes related to energy strings. (New)

266. Specifically, every energy level in the schematic, whether absorption or emission, corresponds to the total amount of internal energy within the electron system. This total internal energy of the electron consists of:
  d. The total number of electrical energy strings
  e. The total number of energy magnetic energy strings
  f. The thickness and lengths of each of those strings

Taken together, the number of energy strings (and their dimensions) in the electron system will give us the total energy of the electron system.
(New)

267. In the absorption schematic, each line represents the new energy of the electron after the incoming photon has been absorbed.

In terms of energy strings we can understand this as follows: When the electron absorbs a photon, the energy strings of the photon enter the electron system. The total amount of energy strings within the electron system has increased. This amount of energy is represented by one of the lines in the schematic. (New)

> *This is my way of showing how the energy level schematic corresponds to the physical reality of energy strings in the absorption process.

268. Each energy level in the schematic for absorption represents a different amount of energy absorbed. This corresponds to different energies (and frequencies) of photons being absorbed. Specifically, this corresponds to the number of energy strings being absorbed.

> (Partially New: schematic is traditional, adding strings is new)

269. The schematic for subsequent *emission* of an EM burst can similarly be related to the processes of energy strings.

When an EM burst is launched, the remaining energy strings are much fewer, resulting in less internal energy of the electron. This is represented by a lower level line in the schematic. (Partially New)

270. The various options of energy levels in the emission schematic can be understood in terms of energy strings and Threshold Percentage.

Whenever a particular group of energy strings merges together in such a way for those string to reach the Threshold Percentage, then those strings will launch as an EM burst. Yet there are several combinations of energy strings which may provide enough energy to launch. Therefore, whichever of those combinations of energy strings is merged together first, then that is what will be launched from the electron.

Thus, every "arrow down" on the emission schematic represents a specific group of energy strings merging and being launched.

And consequently, every line on the emission schematic represents a new internal energy of the electron, after emitting a particular EM burst.

(New)

*This description explains the actual process in terms of energy strings and Threshold Percentage. Therefore this explanation provides a much deeper understanding of the science behind the traditional schematic.

271. Scientists have often wondered how an electron "knows" how to emit a particular frequency of EM burst, in order to drop down to a particular level of energy. We can now explain this fully.

a. The statement is inverted. The actual process is the particular EM burst being emitted first, then the energy of the electron drops to its new energy level.

b. The specific EM burst emitted will depend on which of the possible combinations of energy strings will be created first. This is somewhat random, due to the variety of motions inside the electron system.

Therefore, the electron does not need "know" how to emit a specific frequency in order to arrive at a particular energy state. Rather, the electron emits a particular EM burst depending on which the possible combinations of energy strings is created first. The remaining amount of energy strings in the electron are then the new energy of the electron. Thus, the energy level of the electron "drops" from a higher energy level to another, depending on which group of energy strings is launched first.
(New)
*This is a new clarification for science. This really clarifies the questions of "how does an electron know which frequency to emit" and "which energy state to drop down to" into a more accurate and logical format.

This is significant because scientists often ask how the electron "knows" which level to drop down to.

Also, the reverse of cause and effect is significant: it is not that they know what level to drop down to, it is the particular combination of energy strings, which emits a particular burst, which THEN causes the energy of EM burst to be emitted, and thus results in the electron dropping down to a particular energy level.

Energy Percentage Determines
Which EM Burst (of possible options) will be Created

272. As stated above, most electrons will be able to emit several frequencies of EM bursts. The particular frequency to be emitted (of the existing options) depends primarily on which of the possible launching combinations of energy strings will be created first.

273. The choice of which frequency of EM burst will be launched first can also be related to Threshold Percentage.

For example, when an electron absorbs a high energy photon, the total energy strings can group together in various ways. If the energy percentage is 60%, then one group of energy strings will be emitted, and thus one particular frequency emitted. Yet if 70% of the energy is grouped together and emitted, then a different frequency will be emitted. And if 90% were grouped together, then the third option of frequency would be emitted. (New)

*This is all part of my new ideas, and is an extension of the Threshold Percentage. The grouping of energy strings, as the percentage of energy related to the whole electron system, is what will determine which frequency EM burst will be emitted. This is one of my new ideas, and explains things well.

Emitting the Possible EM Bursts
Starting from a Different Level of Energy

274. In our excited electron system, we can start from any one level of energy, to drop down to any lower level of energy.

275. No matter where we start from, the concept is the same: the exact amount of energy strings grouped together, to reach the next Threshold Percentage, is the determining factor in what frequency will be emitted.

However, the calculations to determine the Threshold Percentage will be different, because it depends on how much energy was absorbed in the beginning. (New)

Emitting a Series of Sequential EM Bursts
Starting from a Different Level of Energy

276. The subsequent EM bursts observed can also be sequential. This occurs when the electron launches one EM burst, yet this electron still has enough energy to launch a second EM burst. This will occur as follows:

   a. The electron absorbs an incoming photon

   b. The new energy reaches the highest level (such as level 4)

   c. The energy strings arrange to emit one EM burst: For example, this EM burst represents the stages of energy levels from level 4 to level 3.

   d. If there is enough energy left over, then remaining energy strings will combine, and have enough energy to break free from the electron. Thus a second EM burst is emitted. For example, this energy of EM burst would represent the change from Level 3 to Level 2.

   e. The process can occur again. For example, the remaining energy strings can combine and launch an EM burst. In this case the energy emitted represents the change from level 2 to level 1.

   f. This process will occur until there are not enough remaining strings to combine, or the remaining strings when combined will never become strong enough to break free from the electron. At this point, we have a stable electron, which will never emit another EM burst.

(Partially New)
*This general concept has been known about for a long time. I have merely added the energy strings.

277. If EM bursts are emitted sequentially, then the energies of each EM bursts added together would be the same as a greater energy EM burst emitted at one time.

278. Calculating the Threshold Percentage for each of several options of EM bursts will first depend on the Starting Energy of the system.

Remember that the amount of energy required to emit each frequency (energy) of EM burst will always be the same (for each option) regardless of where we start. However, the Threshold Percentage depends not only on the amount of energy to launch the EM burst, but also on the amount of energy we begin with.

279. Calculating the Threshold Percentage for each possible option of EM burst is done as follows:

We begin by knowing the total energy of the system. This is the additional energy added (usually this is the energy of the photon absorbed) plus the original amount of internal energy.

Then we look at the possible energy string combinations which will create launch of specific EM bursts. This requires two values: a) the energy of the EM burst which will be emitted, and b) the energy required to break that group of strings away from the electron.

Finally, for each possible EM burst to be emitted, we compare the amount of energy required to launch that EM burst to the total amount of energy in the system. This is the Threshold Percentage to launch that particular energy (frequency) of EM burst from the electron.

We then do this for each of the possible EM bursts to be emitted. From this, we can see the Threshold Percentages for all the possible EM bursts, as related to a particular starting energy.

280. Calculating the Threshold Percentage starting from different levels, or calculating the Threshold Percentage for EM bursts in succession is similar the main Threshold Percentage calculation. However, the calculation is slightly more complex.

We begin with the basic process above, for each possible EM burst, and starting from each possible total energy.

In addition, when we have a series of subsequent emissions, the Threshold Percentage for the second EM burst must be based on the total energy as starting from the new level of energy (after the *first* EM emission).

This is really not as complex as it sounds, but must be considered if you are going to work with these values for predictive purposes.

Chapter 10: Electrons as Particles and Waves
Summary of Concepts

Overview

281. Electrons can be observed to be both particles and waves.

282. Electrons are primarily particles. However, the oscillating energy fields or the vibrating particle motions will make them appear as waves.
    (New)
    *These are the basic ideas. Totally new concepts.

Two Types of Electrons

283. There are two types of electrons:
    a. Atomic electrons (electrons traveling in orbitals)
    b. Free electrons (electrons traveling through the air)

284. Distinguishing between the two types of electrons is important because
    a. These electrons behave very differently. (New)
    b. The wave patterns are created differently. (New)

285. We cannot assume that any observation of a free electron will be the same in an atomic electron. In fact, due to the structure of the atom, it is not possible for the electron in an atom to behave in the same way as a free electron. (New)
    *It is very important that we are very careful about assumptions we make about atomic electrons based on observations of free electrons. Such assumptions can lead us into false answers.

286. The general nature of the atomic electron is as follows: The electron is being pulled inward by gravitational pull. Yet at the same time the electron has internal energy which drives the electron forward.

These combinations of motions keep the electron in an "orbit" or "orbital", where the electron is generally held a certain distance from the nucleus, and yet the electron can freely move around the nucleus.

(Further details of the structure of the electron and the causes of its motions are discussed in a later chapter. Further details of the gravity vs. internal energy relationship will be described in subsequent book on Gravity).

287. The wave patterns produced by the atomic orbital are primarily produced by the oscillating energy waves as the electron travels its path.
(New)
*This is a very new idea. This provides the solution for why electrons appear as waves in an orbital.

288. The energy fields and the wave patterns of the atomic electrons are important for the following processes: emission of EM bursts; absorption of EM bursts; and standing waves in orbiting electrons.
(Partially New)
*The explanation of the wave patterns is new. Showing how the causes of these wave patterns contribute to absorption, emission is new. Showing how the Standing Waves can be explained by this oscillating energy field wave is also new.

289. Free electrons are electrons which are not attached to any atoms. These electrons are forcibly ejected from atoms, at which point the electrons become free to fly through the air. The general nature of the free electron is like a baseball being thrown in the air.

290. Most free electrons are emitted from a device a beam, where electrons are ejected from atoms inside the device, then sent through the opening, in a continuous series of electrons.

291. Usually multiple electrons are ejected at the same time. It is difficult for a beam to produce just "one" electron at a time.

292. The wave pattern of the free electron is created due to the physical motions of the particle itself. The electron spins as it always does. Yet without the countering forces of the nucleus, the spin causes the electron to moves up and down as it travels forward. This results in the electron traveling in a wave pattern. (New)
   *This is the first time that the wave pattern of a free electron has been explained.

293. The wave pattern of the free electron is most important when we discuss the interference patterns created by two or more wave patterns of free electrons.

294. The electron was first determined to be a wave because of experiments on the *free electron* which showed interference patterns.
   However, the results and interpretation of these experiments cannot be applied to the electrons in the atom because the electrons behave differently in the atom than as free electrons. (Partially new)
   *The first paragraph is fact. The second paragraph is my emphasis, based on what I know now about the electron.

Atomic Electron as Particle and Wave: Overview

295. The electron is primarily a particle. It is composed of electric energy strings and magnetic energy strings. The electrical energy strings are what we commonly know of as the electrical field.
   (New – mostly repeated from earlier sections)

296. When an electron travels in an orbital, the energy strings will change direction, depending on where the electron is as compared to the nucleus. (New)
   *This is very new, and will be elaborated on in later books.

297. As the electrical energy strings change direction we observe this as a change in the electrical energy field. The pattern, in total, of this changing electrical energy field is what we observe as a wave in the orbital.
   (New)
   *This basic concept combines Schrödinger's views of the orbital with Bohr's views of the orbital. These are two very different views of the orbital, and yet I have shown how they can be the same.

Electron as Particle and Wave on Power Lines

298. Because the elliptical orbital very similar to alternating current in a power line, we can first understand the electron as particle and wave on the power line, and then extend the concepts to the orbital.

299. In a power line we have alternating current and alternating energy fields. For example, as the electron moves left to right, the electric field may extend upward. Then as the electron moves right to left, the electric field would extend downward.
   Note that the particular direction of the energy field can also extend horizontally, outward from the viewer and inward toward the viewer. All that matters is that the electrical current is perpendicular to the current.

300. The electrical energy field will generally extend the same direction, for as long as the electron is traveling in that direction.
   However, when the electron slows down, and travels in the opposite direction, the electrical field will shrink down, and then extend the opposite direction. (This will be more fully understood when we discuss the model of the electron in a later chapter).
   (New)

301. Therefore, as the electron travels back and forth along the wire as alternating current, the electrical energy field extends up, retreats, and extends down again. This process continues repeatedly, for as long as the alternating current is being produced.

302. We can draw the diagrams of this event:
   a. First as field direction and strength vs electron position
   b. Then as field direction and strength, as time progresses.
   c. Repeating the diagram above, we can see the wave pattern.

   (New)
   *This set of diagrams is very new to the world. This explanation, and the diagrams, simply shows how the oscillating energy fields from the electron motion will create the wave pattern.

303. Therefore, the Wave Pattern is observed with alternating current along a power line. This wave pattern is created by the oscillating electrical energy fields, which accompany the electron as it changes direction. (New)
   *This is the explanation that is new to science.

Electron as Particle and Wave in Orbitals

304. The process in which an electron is both a particle and a wave in an orbital can be seen as an extension of the particle and wave process for an electron on a wire.
   The basic wave pattern of an electron in an orbital is created by the oscillating electrical energy fields as the electron travels its path in the orbital. (New)
   *This summarizes the basic cause of the wave in the orbital. This is a very new concept.

305. The simplest orbital is an ellipse. When an electron travels on an ellipse, it creates a type of alternating current which is similar to that produced on a power line.
   Consequently, the electrical energy fields oscillate in the same way as on the power line. Therefore, the oscillating energy fields as the electron travels the orbital will produce the wave pattern, much as the energy fields did for the electron on the power line. (New)

306. For an electron traveling along an elliptical orbital path, the wave pattern is produced in the following way:

   a. The electron is a particle which travels a continuous elliptical path.

   b. When an electron travels in an elliptical orbit, we will have two types of cycles: alternating current and alternating energy fields.

   c. Because the electron continuously travels this path, these alternating energy fields will continue repeatedly as the electron continues to travel the loop.

   d. If we track the electrical energy fields through moments in time, we can see that the cyclic electrical fields produce a wave pattern of energy.

   (New)
   *This is the full process of how oscillating energy fields produce the wave pattern in an orbital. This process is understood and presented for the first time in this book.

Electric Fields Always Face Nucleus

307. More accurately: for an electron traveling along an orbital, the energy fields of that electron will always extend toward the nucleus. Therefore, rather than the fields extending "up" and "down", the energy fields extend "inward" from side A, then "inward" from side B.
   (New)
   *This is a very new idea.

308. Thus, for an electron traveling along an elliptical orbital path, the wave pattern associated with the electron is *more accurately* produced as follows:
   a. The electron travels around the ellipse.

   b. When the electron travels left to right (side A), the electrical energy field *extends inward, toward the nucleus*.

   c. When the electron rounds the bend, the energy field shrinks (as the field prepares to exit through the opposite side of the electron, and extend through to the other side).

   d. When the electron travels right to left (side B), the electrical energy field *again extends inward toward the nucleus*, yet *from the opposite side* of the orbital, and the opposite side of the electron. Thus, the direction of the electrical energy field is exactly opposite.

   e. As stated above, because the electron continuously travels this path, these alternating energy fields will continue repeatedly as the electron continues to travel the loop.

   f. If we track the electrical energy fields through moments in time, we can see that the cyclic electrical fields produce a wave pattern of energy.

   (New)
   *This is a more accurate understanding of the electron's energy fields, and the creation of the wave pattern of oscillating fields in the orbital. All of this is quite new.

Complex Orbital Shapes and Complex Wave Patterns

309. We can apply all of the concepts described above to any orbital, regardless of how complex it is.

    The orbital shape can be very complex, with many twists and turns, yet the electrical energy fields will extend, retract, and extend in the opposite direction (depending on the particular direction the electron is traveling). This will produce oscillating electrical energy fields; the pattern may be complex, but it will be a wave pattern, and it will repeat.

    Therefore, any complex path of the electron, in any complex orbital shape, will create a complex wave pattern of electrical energy.

    (New)

    *This is simply extension of all the concepts described above. However, it is important to state for clarity that these concepts can be applied to any orbital, of any shape.

Standing Waves of Electrons

310. A "standing wave" is a wave pattern that is confined to a particular region of space.

311. A standing wave will only exist if the following conditions are met:
   a. The wave is repeated continuously
   b. The wave is repeated in a specific region of space, and
   c. The wave is the exact same wave pattern, no matter how complex that pattern is.

   (Partially New)
   *This is my list of requirements. I am sure there are other lists.

312. In an orbital, a standing wave is created by the repeated oscillations of energy fields. Specifically:

   a. The electron travels a path which is a closed loop. This is what we know as the orbital.

   b. The direction of the energy fields depend on where the electron is located in relation to the nucleus.

   c. Therefore the energy fields change direction as the electron changes direction.

   d. This produces an oscillation of energy fields as the electron travels along its path. This is what we know of as the electrical energy wave.

   e. Then because the electron travels the same path repeatedly, the specific wave pattern is produced repeatedly.

   f. Therefore, the same wave pattern of oscillating electrical energy is produced, repeatedly. In other words, we have created a "standing wave."

   Thus, the same oscillating energy field is produced repeatedly, which produces the same energy wave pattern repeatedly. This is the same oscillations, in the same directions at the same locations. By definition, this is exactly the formation of a standing wave. (New)
   *This is very new. Physicists have talked about the idea of a "standing wave" associated with the orbital, yet no one knew exactly what this wave would be, or how it was created. This is the first time that anyone has explained the standing wave of the electron in the orbital.

313. A standing wave can be extremely complex. As long as the three requirements above are met, the electron can create a standing wave.

The electron may have numerous twists and turns, the electron may veer in any direction, at any angle. A standing wave will always be created.

This is because every time the electron changes direction, the energy field also changes direction. Therefore, the oscillating wave of energy field will be created. The pattern may be quite complex, yet the wave pattern will exist. And because the electron travels the same complex path repeatedly, the energy fields will create the same complex wave pattern.

Therefore, regardless of how complex the path is, the complex wave pattern will repeat forever. Thus, a standing wave will always be created in an orbital. (New)

*These concepts are natural extensions of concepts discussed above.

Electron as Blurred Energy

314. The electron in an orbital is often considered to be blurred energy. We can now understand this as the standing wave of oscillating electrical energy fields. The appearance of the blurred electron is in fact the electrical energy field which oscillates as the electron travels its path around the orbital. (New)

*This is the first time anyone has shown exactly how an electron is blurred energy; while still being primarily a particle.

315. The "smeared" electron is an illusion. In reality, the energy field extends in particular directions at particular locations (based on where the electron is at any given moment). Yet this process is much faster than our devices can detect. Therefore we observe only a "blur" of energy.

The electron appears to be everywhere at once, and the energy waves appear to be everywhere at once. Yet in fact it is due to the speed of the oscillating energy waves in relation to our observation devices, along with fact that the path is repeated (the standing wave pattern) that we detect a blur of energy as if everything happened simultaneously.

(New)

*This clarifies why the electron can appear as a blur, though in fact it is simply energy which moves very quickly. Also, we are making a distinction between reality and what our devices can detect, which then causes the difference between actual motion vs. apparent motion.

Creating the Electron Wave for a Free Particle

316. The process for creating the wave pattern for the free electron is different than the process for the wave pattern in an atomic electron.
    (New)
    *It is very important to make this distinction. The processes are very different.

317. The wave pattern of the free electron is created as a combination of motions of the particle itself, rather than involving the fields.

318. The process of creating the wave pattern for a free electron is very similar to the process of creating the wave pattern for the photon.
    There are two motions: 1) up and down motion, and 2) forward motion. The combination of these motions will create a wave pattern as the electron flies through the air. (New)
    *This is significant. I am the first person to really explain how the free electron creates a wave pattern as it travels.

319. More specifically, the electron has internal energy. This internal energy causes all motions, including: spin, vibration, and forward motion. (The details will be explained and illustrated in a later chapter). Therefore, the internal energy of the free electron will create a combination of motions. The net result of these motions is to produce a wave pattern as it travels forward. (New)
    *This is significant. I am the first person to really explain how the free electron creates a wave pattern as it travels.

Two Methods of Electron as Particle and Wave Reviewed

320. There are two processes for an electron to create a wave pattern. In brief these are:
   a. The energy field oscillates, thereby creating a wave pattern.
   b. The particle vibrates and moves forward, thereby creating a wave pattern.
   (New)
   *This is reviewing from concepts above. Yet these concepts are very important.

321. The two processes occur in two different environments. The difference is due to the effect of the nucleus.

   a. In the atomic orbital, the wave pattern is created by oscillating energy fields as the electron particle travels around the nucleus.

   b. As a free electron, the wave pattern is created by the interior energy causing the particle to vibrate and move forward, thus creating the wave pattern.

   (New)
   *Again, this is reviewing from concepts above. Yet these concepts are very important.

322. The wave pattern is not produced in the same way by the other type of electron because of the presence or absence of the nucleus. The nucleus provides a set of forces which influences the behavior of the electron and the energy fields. (New)

323. In the free electron, the energy field does *not* oscillate, simply because there is no nucleus. The energy fields have no reason to aim one direction over another. Therefore in a free electron the energy fields essentially remain fixed in place while the electron itself vibrates and rotates. (New)

324. However, the energy fields do "change direction" in the sense that as the electron spins, the fixed energy fields also rotate (as in a set of needles on a ball). This will produce some change in energy fields, and may in fact produce an oscillating pattern, yet not the same process, or the same degree, as in the atomic orbital. (New)

325. In a free electron, the up and down motion of the electron is more pronounced, primarily because the free electron has more internal energy.

In order to become a free electron this electron must have enough energy to break away from the gravitational pull of the nucleus. Therefore, most free electrons will have more internal energy than most atomic electrons (depending which atoms to which we are comparing the free electron).

Therefore, the free electron has more internal energy. This means it will vibrate up and down to a larger extent. (It will also possibly travel forward at a faster rate). If the electron vibrates up and down to a larger extent, then the wave pattern becomes much more noticeable.

(New)

326. In the atomic electron, the electron does *not* move up and down as much as the free electron does because of: 1) the internal energy of the electron, and 2) the gravitational pull of the nucleus.

Restated from above (in the opposite way) the internal energy of the atomic electron is usually less than the internal energy of the free electron. This is because the atomic electron does not yet have enough energy to break free from the nucleus.

And since the atomic electron does not have as much internal energy, the vibrational motion will be less. Therefore, the up-down motion of the atomic electron will be much less significant, if it exists at all. (New)

General Principle for Particle-Wave Duality

327. Other particles such as the free proton can also create a wave pattern. This will occur in the same way as the free electron:
   a. The proton is free from the nucleus
   b. The proton vibrates up and down
   c. The proton travels forward.

The combination of the up-down motion and the forward motion for the free proton will create a wave pattern as the proton travels. (New)

328. The internal mechanism for the proton is similar to the internal mechanism of the electron. Both are particles, with magnetic energy inside. The magnetic energy drives the motions of the proton (the same as with the electron), thereby producing spin, vibration, and forward motion. (This will be explained and illustrated in other books and papers).
   Together, these motions will create a wave pattern as the proton flies forward. (New)

329. It is important to remember that only the free proton will travel as a wave. The proton in the nucleus will not create a wave.

   a. <u>The proton in the nucleus cannot make a wave by particle motion</u>.
   The proton in the nucleus does not move. It essentially fixed in place, held to the other protons and the neutrons by the nuclear force. Therefore the proton in the nucleus cannot move in a wave pattern.

   b. <u>Similarly, the proton cannot make a wave by energy fields</u>.
   First, the electrical energy field does not extend outward. (See my paper: "Quarks Simplified and a Unified Field Theory" for further details). Therefore, without the electrical field extending outward, we cannot have the wave due to energy fields. Therefore, the proton in the nucleus will never create a wave pattern.
   (New)
   *This is important to note, just for understanding new details of the atom, the proton, and the electron, which are discussed throughout my books and papers.

330. **The General Principle of Particle-Wave Duality #1**:
   Any particle flying through the air, regardless of whether or not it has an energy field, can be tracked as a wave, if it:

   1. Is a Free Particle
   2. Oscillates Up and Down
       This includes: pulsating, vibrating, or migrating;
       in one direction and then the other, repeatedly.
           AND
   3. Travels Forward at the same time.

   This combination of motions will produce a wave pattern, for any particle; thereby creating Particle-Wave Duality

   (New)
   *This is a General Principle of Particle-Wave Duality. It is a very big statement. I am the first person to do three things: I am the first person to solve the mystery of particle-wave duality for electromagnetic energy. I am the first person to solve the mystery of particle-wave duality for free subatomic particles such as the electron and the proton.
   Then, putting it all together, I am the first person to create a General Principle which summarizes all of particle-wave duality, for any particle which flows through the air.
   Yes, in total, this General Principle of Particle-Wave Duality is a huge contribution to the scientific world. And you read it here first!

   *Note that this statement will also explain many wave patterns created by water molecules and other particles.

331. The General Principle of Particle-Wave Duality, #2:
Any particle can also be tracked as a wave, if it:

1. Has an oscillating *energy field*
   AND
2a. The particle travels in a continuous loop (such as an orbital)
   OR
2b. The particle travels in a forward motion through the air

This combination of factors will produce an oscillating energy field, which repeats continuously, while the particle moves ahead, and can then be tracked as a wave.

(New)
*This General Principle uses the Energy Fields rather than the particle motions. Thus, this General Principle applies to electron orbits, as well as any other particles with significantly noticeable oscillating energy fields.

This General Principle is also very new to the world, and is a significant contribution.

332. The wave pattern created by electromagnetic energy can be placed in either of these two General Principles.

In a later book we will show exactly how the pulsation of electromagnetic energy is created. In brief: the energy field strings expand outward, then come back inward, at which point they proceed outward again. This of course is done while the photon travels forward.

Therefore, this process can be viewed as the First General Principle, where the strings "migrate" in an oscillating pattern (while the particle moves forward). Yet this can also be viewed as the Second General Principle, where the energy field itself is oscillating (while the particle) moves forward.

(New)
*Just a few final thoughts on understanding electromagnetic energy.

Made in the USA
Columbia, SC
04 March 2021